United States Nuclear Regulatory Commission

Protecting People and the Environment

Proceedings of the

Workshop on Probabilistic Flood Hazard Assessment (PFHA)

U.S. NRC Headquarters
Rockville, Maryland

January 29-31, 2013

AVAILABILITY OF REFERENCE MATERIALS
IN NRC PUBLICATIONS

U.S.NRC

United States Nuclear Regulatory Commission

Protecting People and the Environment

Proceedings of the

Workshop on Probabilistic Flood Hazard Assessment (PFHA)

U.S. NRC Headquarters Rockville, Maryland

January 29-31, 2013

Manuscript Completed: June 2013

Date Published: October 2013

Prepared by: T. J. Nicholson and W. A. Reed

DISCLAIMER

This report is not a substitute for U.S. Government regulations, and compliance with the information and guidance provided is not required. The technical approaches, software, and methods described in these conference proceedings are provided for information only. Publication of these proceedings does not necessarily constitute Federal agency approval or agreement with the information contained herein. Use of product or trade names is for identification purposes only and does not constitute endorsement or recommendation for use by any Federal agency.

The views expressed in these proceedings are those of the individual authors and do not necessarily reflect the views or policies of the U.S. Nuclear Regulatory Commission and the other participating Federal agencies, but those of the USGS-authors do represent the views of the USGS.

The USGS-authored abstracts contained herein have been peer reviewed and approved for publication consistent with USGS Fundamental Science Practices (http://pubs.usgs.gov/cir/1367/).

ABSTRACT

The U.S. Nuclear Regulatory Commission's (NRC's) Offices of Nuclear Regulatory Research (RES), Nuclear Reactor Regulation (NRR) and New Reactors (NRO) organized this *Workshop on Probabilistic Flood Hazard Assessment (PFHA)*. The workshop was held January 29–31, 2013 at the NRC headquarters auditorium, 11545 Rockville Pike, Rockville, MD. The workshop was coordinated with Federal agency partners: the U.S. Department of Energy (DOE), U.S. Department of the Interior's Bureau of Reclamation (BoR) and U.S. Geological Survey (USGS), U.S. Army Corps of Engineers (USACE), and Federal Energy Regulatory Commission (FERC). The research workshop was devoted to the sharing of information on probabilistic flood-hazard assessments for extreme events (i.e., annual exceedance probabilities much less than 2.0E–3 per year) from the Federal community. The topics included: Federal agencies' interests and needs in PFHA; State of the Practice in Identifying Extreme Flood Hazards; Extreme Precipitation Events; Flood-Induced Dam and Levee Failures; Tsunami Flooding; Riverine Flooding; Extreme Storm Surges for Coastal Areas; and Combined Events Flooding. The workshop objectives included to: (1) assess, discuss, and inform workshop participants on the state of the practice for extreme flood assessments within a risk context, (2) facilitate the sharing of information between both Federal agencies and other interested parties to bridge the current state of knowledge between extreme flood assessments and risk assessments of critical infrastructures, (3) seek ideas and insights on possible ways to develop a PFHA for use in probabilistic risk assessments, (4) identify potential components of flood-causing mechanisms that lend themselves to probabilistic analysis and warrant collaborative study, and (5) establish realistic plans for coordination of PFHA research studies. Observations and insights provided during session presentations and subsequent panel discussions that followed were documented by the panel reporters and are included in this report.

FOREWORD

The U.S. Nuclear Regulatory Commission (NRC) in cooperation with its Federal partners—the U.S. Department of Energy (DOE), U.S. Bureau of Reclamation, U.S. Army Corps of Engineers, U.S. Geological Survey, and Federal Energy Regulatory Commission—organized and conducted a *Workshop on Probabilistic Flood Hazard Assessment (PFHA)* at NRC headquarters in Rockville, Maryland from January 29–31, 2013. The research workshop was devoted to sharing information on probabilistic flood hazard assessments for extreme natural events such as extreme precipitation events, flood-induced dam and levee failures, tsunamis flooding, riverine flooding, extreme storm surges along coastal areas, and combined events flooding. Other technical contributors to the workshop were the National Weather Service of the National Oceanic and Atmospheric Administration and the Federal Emergency Management Agency and invited industry, academic, and DOE national laboratory experts in the field. Significant technical support was also provided by Deltares, an independent institute in the Netherlands for applied research in the field of water, subsurface, and infrastructures.

The NRC published this proceedings report of the PFHA workshop to further share the knowledge and practical experiences presented at the workshop and to summarize the significant insights and observations identified by the workshop participants. The workshop proceedings provide a glimpse into the state of the practice and science of flood hazard assessment for extreme natural events within a risk context. The report provides references and electronic links to information sources presented and discussed at the workshop. For example, all of the workshop presentation slides, the video of the workshop, the Webcast, and the public meeting summary can be viewed at the NRC public Web site page: http://www.nrc.gov/public-involve/public-meetings/meeting-archives/research-wkshps.html.

We hope this workshop and its proceedings will serve the Federal community and other interested stakeholders who are using or considering a probabilistic approach for flood hazard assessments. Cooperation and collaboration among Federal and local agencies and stakeholders in the sharing of flood risk information is very important to leverage limited resources and to serve the American public. The workshop participants and their technical interactions demonstrated the benefit of such strong cooperation and collaboration.

K. Steven West, Deputy Director
Office of Nuclear Regulatory Research
U.S. Nuclear Regulatory Commission

TABLE OF CONTENTS

FIGURES

TABLES

EXECUTIVE SUMMARY

The U.S. Nuclear Regulatory Commission's (NRC's) Offices of Nuclear Regulatory Research (RES), Nuclear Reactor Regulation (NRR), and New Reactors (NRO), in cooperation with Federal agency partners: U.S. Department of Energy (DOE); Federal Energy Regulatory Commission (FERC); U.S. Army Corps of Engineers (USACE); U.S. Bureau of Reclamation (BoR); and U.S. Geological Survey (USGS) developed a research Workshop on Probabilistic Flood Hazard Assessment (PFHA). The motivation for holding the workshop was to provide a forum for Federal agencies and other organizations to exchange information regarding flood hazard assessments, and to establish the state-of-the-practice in terms of such assessments within a risk context. Flood assessments include combinations of flood-causing mechanisms associated with dam and levee safety, probable maximum floods, probable maximum precipitation, riverine flooding, hurricane storm surge, and tsunami.

The objectives of the workshop were to (1) identify and solicit presentations on the state-of-the-practice in extreme flood assessments within a risk context, (2) facilitate the sharing of information to bridge the current state-of-knowledge between extreme flood assessments and risk assessments of critical infrastructures, (3) seek ideas and insights on possible ways to develop a probabilistic flood hazard assessment (PFHA) for use in probabilistic risk assessments (PRAs), (4) identify potential components of flood-causing mechanisms that lend themselves to probabilistic analysis and warrant further study (e.g., computer-generated storm events), and (5) develop plans for a cooperative research strategy on PFHA for the workshop partners.

An organizing committee composed of Federal representatives and academic and foreign experts was established (please see Acknowledgments for a listing of members) to identify the flood-causing events that were used to formulate the panel session themes. The committee also recommended and elicited participation from appropriate experts as invited presenters and panelists. These experts included Federal staff, academic experts, and representatives from Deltares, an independent research institute for water management issues based in the Netherlands. The committee developed an agenda based upon the identified panel themes, and chose the panel co-chairs. Panel co-chairs developed a series of questions to help focus the panel discussions. Technical reporters captured significant insights and recommendations that were used to develop these proceedings.

Prior to the panel sessions, NRC Commissioner George Apostolakis provided the keynote address that set the stage for the rest of the workshop. His presentation outlined the importance of risk-informed, performance-based regulations and highlighted the important interplay between probabilistic assessments and traditional deterministic methods.

Panel Summaries

Panel 1 focused on the participating Federal agencies' interests and needs regarding PFHA. Presentations from a number of Federal agencies highlighted their efforts in developing, using, and implementing probabilistic methods for flood assessments. These efforts ranged from agencies already using PFHA in a risk-informed framework (e.g., BoR) to those agencies in the planning stages of introducing probabilistic methods (e.g., FERC).

Representatives of the American Nuclear Society (ANS) working groups responsible for the revision of the flood-related standards ANSI/ANS-2.8, "Determining External Flood Hazards for Nuclear Facilities," and ANSI/ANS-2.31, "Determining Design Basis On-site Flooding Caused by Precipitation at Nuclear Facility Sites," updated the participants on the progress of the relevant working groups and how these standards are incorporating probabilistic principles. The proceeding discussion focused on determining the impediments to implementing PFHA, how to address uncertainties, and the synergy of deterministic and probabilistic analysis.

Panel 2 focused on the state-of-the-practice in identifying and quantifying extreme flood hazards including their frequency and associated flood conditions within a risk context. Presentations covered the history of flood frequency analysis; the basic concepts, philosophy, and strategy underlying current analytical methods for extreme floods; the uses of paleoflood data to provide information on large infrequent floods; and methods currently used by Federal agencies for flood hazard analysis.

The main focus of the discussion period was on uncertainties in the data used for quantifying extreme flood events, how paleoflood data can be used to improve the historical data record, and the impact of climate change on flood estimations. Similar to the Panel 1 session, Panel 2 also discussed the hindrances to using PFHA.

Panel 3 examined extreme precipitation events and their impacts on flooding due to local or watershed-scale responses. The presentations covered topics such as the use of radar to measure rainfall and its advantages over traditional rain gauges, the use of modeling of extreme precipitation events, and the determination of extreme precipitation frequencies. During the discussions, the panelists expanded on presentation themes and discussed the advancements made in extreme precipitation data over the past 30 years. Radar rainfall measurements were used as an example because they provide a better picture of spatial and temporal correlation of rainfall data for models. However, these measurements have limitations in that rainfall for intense events can be underestimated. The panelists also discussed the advances in statistical processing methods (e.g., frequency estimation) as well as developments in physical and numerical modeling of extreme precipitation (e.g., the Weather Research and Forecasting Model, WRF).

Panel 4 focused on flood-induced dam and levee failure. The first presentation provided an overview of risk assessment for dam and levee failures and stressed the importance of thorough identification of hazards. It also introduced the concept of tolerable risk. The next presentation illustrated the Dutch approach to dam and levee failure—since 1953, the Dutch have used risk-based analysis and a risk-based approach to determine storm surge protection barriers. FERC's first foray into risk-informed decision-making was described in relation to the determination of the Inflow Design Flood for two high-hazard dams. The BoR's method of calculating dam breaches was described, and an overview of the USACE dam and levee safety program, which uses a risk analysis approach, was presented. The panel discussion determined a number of insights and observations that included the need to improve risk assessment procedures within Federal agencies (e.g., by improving failure mode identification criteria); better scoping of risk assessments and risk model formulation; better consideration of uncertainties; and better understanding of the time scale of flood causing and dam failure events. It was also mentioned that no good databases exist for dam failures, especially for large dam breaches, and that reliable data for dam components are not available. The need for a Senior Seismic Hazard Analysis Committee (SSHAC)-like process is needed because specific analysis often relies upon expert assessments.

Panel 5 addressed tsunami flooding with a focus on Probabilistic Tsunami Hazard Assessment (PTHA), which is derived from Probabilistic Seismic Hazard Assessment (PSHA). The first set of presentations addressed the fundamentals of viability of extending the analysis to extreme probabilities ($10^{-4} - 10^{-6}$). Other presentations discussed how aleatory and epistemic uncertainties are incorporated and also focused on landslide tsunamis, the procedures for developing landslide size and occurrence distributions, and hydrodynamic methods used to simulate tsunami propagation and inundation. During the discussion session, the presenters stressed the importance of using propagation and inundation models that have been well tested. In addition, they identified the important input parameters for such models, which included source-size distribution, and recurrence rate and distribution. The panel also examined the uncertainties related to these parameters, including a discussion of the methods and data available to test hypotheses and distributions. A major observation was that more work needs to be done in compiling tsunami source parameters.

Panel 6 focused on riverine flooding and included precipitation events and antecedent events such as snow pack melt. This session featured five presentations that included an overview of the current status of PFHA at the NRC and its associated guidance for assessing flood hazards.

The presentations described how the USACE developed unregulated and regulated frequency curves on the Missouri River as well as how ratios of observed floods were used to extend the regulated frequency curves. In addition, the Storm Event Flood Model (SEFM) was described, including how Monte Carlo simulation can be used to simulate thousands of years of peak flood data. The BoR presented a practical use of the SEFM and paleoflood data for estimating flood discharges up to exceedance probability of 10^{-5} on the North Fork Red River near Altus Dam in Oklahoma. Finally, the BoR provided further description of how it uses paleoflood data to evaluate flood hazards at BoR dams. Presenters spent much of the discussion period debating how more data can be obtained for models such as the SEFM model, which can be used to estimate extreme (10^{-5} or 10^{-6}) flood discharges.

Panel 7 looked at extreme storm surge for coastal areas. Extreme events such as Hurricane Katrina and Superstorm Sandy highlighted the need to be able to model and establish frequency of storm surge with regards to design of coastal facilities and adequate emergency planning. The first presentation provided a description of the Dutch approach to coastal flood hazard assessment and the requirements that a periodic safety assessment is carried out every 5 years. Also, a description was presented concerning the statistics used to inform the failure models that are used within the risk assessments such as river discharge and soil cohesion. The presentations included a description of the National Oceanic and Atmospheric Administration's (NOAA) use of storm surge modeling in its mission as well as models such as SLOSH - Sea, Lake and Overland Surges from Hurricanes model, which uses storm track radius of hurricane winds and central pressure, and ETSS - Extratropical Storm Surge model, which predicts storm surge heights. In addition, the final presentation described the Federal Emergency Management Agency's coastal flood hazard analysis framework and the USACE's modeling framework that is used to look at very low probability events and flood responses. This final discussion introduced a joint effort between a number of Federal agencies, the Earth System Modeling Framework (ESMF), of which the USACE's CSTORM-MS is part. The final speaker presented an academic perspective of storm surge studies and highlighted the unique attributes of coastal flooding such as the dynamic landscape of the affected area. In addition, the speaker provided an explanation of how storm surge response functions are calculated and how progress has been made in mathematically accounting for aleatory uncertainties. During the discussion period, the final speaker addressed the topic of climate change and sea level rises, and the consensus seemed to be that a need exists for being adaptive in modeling and allowing room for uncertainty. The discussion highlighted that, in the field coastal storm evaluation, both Federal agencies and organizations can contribute different perspectives to the discipline such as digital elevation models from FEMA and the impact of cover changes on storm effects from Virginia Polytechnic Institute and State University (Virginia Tech).

Panel 8, which was the final technical panel session, looked at how site-specific flood hazards can come from combinations of flood-causing mechanisms and how no single prescriptive set of scenarios is adequate as the design basis flood. Presentations included processes that should be considered for a design basis flood. Often, these processes consist of a primary event followed by a secondary coincident or antecedent event. Though these processes are deterministic, experts in the field are now recognizing the limitations of this process and considering the incorporation of probabilistic approaches. USACE described the ways of assessing levee performance and using risk analysis modeling. Models include HEC-FAA, a flood damage reduction analysis, and HEC-WAT, an emerging water-analysis tool that can account for combined events. The session stressed the importance of considering all uncertainties, even if they can't be quantified. Using seismic hazard as a reference point, one presentation explained how the use of logic trees can be used to model uncertainties and account for all alternative data interpretations. Another presentation described development of a PFHA for a tidal river system near Rotterdam, the Netherlands. This analysis took into account combinations of forcing variables such as river discharge and storm surge barrier operations.

This study found that large contributions to risk were obtained from combinations of storm surges and discharges that were much less likely than a single extreme event. The analysis illustrated that dam failures may occur as a combination of various events, both natural and manmade. The panel session discussion provided several insights into understanding the formulation and analysis of combined events. Principal insights were that, although extreme events should be considered, combinations of more frequent events should not be overlooked when performing a safety assessment, and there is a need to share and document both failures of operation and successes, including near-misses, because they would provide valuable insight into risk assessment processes.

Panel 9 provided an opportunity for the co-chairs and technical reporters from the eight technical sessions to summarize their session's observations and insights. In particular, the panel co-chairs made short presentations that highlighted the more significant points identified during the panel presentations and discussions. These points were:

- Risk-informed approaches are being used and incorporated in safety assessments and decision-making by Federal agencies and international groups.

- It is not a question of deterministic versus risk assessment because they are complementary processes. PFHA requires probabilities of initiating events and facilitates uncertainty analysis.

- Consider inclusion of SSHAC-type of approaches for selected hazards to address gaps in data and analytical methods.

- An expert elicitation strategy similar to the SSHAC would help address:
  - paucity of data for characterizing extreme events,
  - formulation of scenarios in hydrometeorologic model simulations, and
  - systematic assessment uncertainties (epistemic and aleatory).
  -

- Researchers are meeting many of the technical challenges to implement PFHA; however, impediments to applying PFHA include lack of data on frequency, magnitude and duration of the events, willingness to try PFHA, availability of experts, and communication of information on statistical and risk assessment methods.

- PFHA strategies need multidisciplinary teams to:
  - assess complex meteorologic, hydrologic, and geologic data;
  - simulate hydrometeorologic conditions and scenarios; and
  - establish and conduct assessments within a risk framework.

- University and Federal training programs need to focus on courses in statistics, risk, and uncertainty assessments.

- Understanding of the commonality and differences in risk-informed approaches and decision criteria among the Federal agencies must be established.

- Collaborative and coordinated efforts with Federal and State agencies, industry, standards bodies (e.g., American National Standards Institute and ASTM), and other stakeholders must be established to develop mutually accessible data bases and models.

Suggested Areas for Further Work

- Develop a systematic process of expert elicitation for flood hazard assessment (EEFHA). The EEFHA would fill information gaps in flood event scenarios for estimating probabilistic flood hazard magnitudes, durations, and frequencies. The EEFHA process should include uncertainty assessments of the flood scenarios, the past history of floods including paleofloods and regional storm events, and related storm event parameters.

- Support the ongoing development of the USACE's Storm Catalogue for analyzing floods in the United States. The catalogue relates extreme storms to flood events and includes both point measurements and radar data for spatial and temporal distribution of the precipitation. This information will provide technical background for both the expert elicitation process and for site-specific stochastic modeling of extreme floods (e.g., Stochastic Event Flood Model).

- Develop a structured evaluation process for dam and levee failures within the EEFHA process to examine comprehensive uncertainties in data and modeling of potential failure mode scenarios.

- Further develop and apply the USACE's joint probability method for storm and hurricane surge analyses along the Gulf and Atlantic coasts with possible application to the Great Lakes.

- Integrate risk analysis into the state-of-the-practice of watershed and coastal-storm surge modeling as presented by the BoR and USACE.

- Support ongoing interagency committee activities such as the Subcommittee on Hydrology's working groups on hydrologic frequency analysis and extreme storm events.

In total, over 250 participants registered for the workshop, including those viewing remotely using the NRC's webstreaming service. The participants represented Federal employees, contractors, industry representatives, academics, and countries such as France and the Netherlands were also represented. A list of all workshop attendees and their affiliations can be found in Appendix B.

The PFHA Workshop Program (please see Appendix A), and all of the presentations can be viewed at:
http://www.nrc.gov/public-involve/public-meetings/meeting-archives/research-wkshps.html.

ACKNOWLEDGMENTS

The concept, planning and execution of this workshop, and the development and documentation of these proceedings were achieved by an organizing committee composed of Federal agency and non-Federal volunteers. The organizing committee consisted of: Tom Nicholson, Richard Raione, and Christopher Cook, NRC, Co-Chairs; John England and Tony Wahl, BoR; Mark Blackburn, DOE; Tony Cheesebrough and Joel Piper, DHS; Siamak Esfandiary, FEMA; Sam Lin and David Lord, FERC; Chandra Pathak, David Margo, and Ty Wamsley, USACE; Timothy Cohn and Eric Geist, USGS; Donald Resio, University of North Florida; Joost Beckers, Deltares; Fernando Ferrante, Joe Kanney, Sunil Weerakkody, Jeff Mitman, Nathan Siu, and Wendy Reed, NRC.

Many of the organizing committee members, and the invited presenters and panelists were members of the Advisory Committee on Water Information's (ACWI) Subcommittee on Hydrology and its work groups.

The organizing committee is grateful for the support of the Panel Co-Chairs:

Joost Beckers, Deltares
Mark Blackburn, DOE
Nilesh Chokshi, NRC
Timothy Cohn, USGS
Christopher Cook, NRC
John England, BoR
Eric Geist, USGS
Henry Jones, NRC
Sam Lin, FERC

Dave Margo, USACE
Thomas Nicholson, NRC
Chandra Pathak, USACE
Rajiv Prasad, PNNL
Don Resio, University of North Florida
Will Thomas, Michael Baker, Jr., Inc.
Ty Wamsley, USACE
Tony Wahl, BoR

and session reporters:

Hosung Ahn, NRC
Michelle Bensi, NRC
Peter Chaput, NRC
Chris Cook, NRC
Randy Fedors, NRC
Fernando Ferrante, NRC
Mark Fuhrmann, NRC
Joseph Giacinto, NRC
Brad Harvey, NRC
Mark McBride, NRC
Jeff Mitman, NRC
Jacob Philip, NRC
Marie Pohida, NRC
Wendy Reed, NRC
Nebiyu Tiruneh, NRC
Juan Uribe, NRC

We are also appreciative of the NRC Program Support Staff: Tracy Osband, IT; Christine Kundrat, AV-Photo Resource; Woody Machalek and Darrin Raaum, NRC Graphics; and NRC Administrative Assistants: Travin Dashiell, Alicia Griggs, Vivian Mills, Karen Richards, Sylvia Woods; and NRC Security: Gary Simpler and Chris Lamb; who greatly facilitated the logistics for this multi-day and extensive workshop.

Finally, we are indebted to all of the presenters and panelists, who are listed in the workshop agenda (please see Appendix A), for their efforts to prepare their presentations and accompanying abstracts, to travel to Rockville, MD, and to present and discuss their work. We are also indebted to the approximately 260 workshop attendees (listed in Appendix B) who actively engaged in questioning the speakers and panelists during the panel discussions. Together these presenters and panelists generated significant information and observations (as noted in the abstracts and panel session summaries, documented in these proceedings), which made for a very successful workshop and proceedings.

INTRODUCTION

1.1 Background to the Probabilistic Flood Hazard Workshop

Following the extreme flood event at the Japanese nuclear power reactor site at Fukushima Dai-ichi, the U.S. Nuclear Regulatory Commission's (NRC's) Offices of Nuclear Regulatory Research (RES), Nuclear Reactor Regulation (NRR), and New Reactors (NRO) identified research needs for understanding and developing a probabilistic approach for extreme flood assessments within a risk context. The NRC technical staff contacted its Federal counterparts on the Federal Advisory Committee on Water Information's (ACWI's) Subcommittee on Hydrology (SOH) to solicit information on the state of the practice in assessing probabilistic flood hazards. Many of these agencies were either developing or interested in pursuing a probabilistic approach for assessing extreme flood hazards. In particular, the SOH workgroup on Hydrologic Frequency Analysis is actively revising Bulletin 17B, "Guidelines for Determining Flood Flow Frequency." The principal Federal agency partners, the U.S. Department of Energy (DOE); Federal Energy Regulatory Commission (FERC); U.S. Army Corps of Engineers (USACE); and U.S. Department of the Interior's Bureau of Reclamation (BoR) and U.S. Geological Survey (USGS) worked together to develop a *Workshop on Probabilistic Flood Hazard Assessment* (PFHA): The National Weather Service (NWS) of the National Oceanic and Atmospheric Administration (NOAA) and the Federal Emergency Management Agency (FEMA) also provided information and workshop presenters. (Please see Appendix C for a listing of acronyms used in these proceedings.)

The motivation for holding the workshop was to provide a forum for Federal agencies and other organizations to exchange information regarding flood hazard assessments and to discuss the state of the practice in terms of such assessments within a risk context. Flood assessments include combinations of flood-causing mechanisms associated with dam and levee safety, probable maximum floods, probable maximum precipitation, riverine flooding, hurricane storm surge, and tsunamis. The focus of the research workshop was on probabilistic flood hazard assessments for extreme events (i.e., annual exceedance probabilities (AEP) much less than 2.0E-3 per year (please see Appendix F for AEPs related to recurrence intervals)).

1.2 PFHA Workshop Objectives

The objectives of the workshop were to (1) identify and solicit presentations on the state of the practice in extreme flood assessments within a risk context, (2) facilitate the sharing of information to bridge the current state of knowledge between extreme flood assessments and risk assessments of critical infrastructures, (3) seek ideas and insights on possible ways to develop a probabilistic flood hazard assessment (PFHA) for use in probabilistic risk assessments (PRAs), (4) identify potential components of flood-causing mechanisms that lend themselves to probabilistic analysis and warrant further study (e.g., computer-generated storm events), and (5) develop plans for a cooperative research strategy on PFHA for the workshop partners.

1.3 PFHA Workshop Agenda and Logistics

An organizing committee composed of Federal representatives and academic and foreign experts was established (please see Acknowledgments for a listing of members) to identify the flood-causing events that were used to formulate the panel session themes. Many of the organizing committee members were from Federal partner agencies, many of whom are members of the SOH and its workgroups. The committee also recommended and elicited participation from appropriate experts as invited presenters and panelists. These experts included Federal staff, academic experts, and representatives from Deltares, an independent research institute for water-management issues based in the Netherlands. The committee developed an agenda based on the identified panel themes, and chose the panel co-chairs (please see Appendix A for the workshop agenda). Panel co-chairs developed a series of questions to help focus the panel discussions. Technical reporters captured significant insights and recommendations that were used to develop these proceedings.

Appendix D provides a bibliography of research papers and information sources used in developing the workshop program.

1.4 PFHA Workshop

The workshop was held January 29–31, 2013, at the U.S. Nuclear Regulatory Commission (NRC) Headquarters Auditorium, 11545 Rockville Pike, Rockville, Maryland. The Workshop was coordinated with the SOH and its member agencies. The workshop technical topics were divided into eight panel sessions:

- Panel 1: Federal Agencies' Interests and Needs in PFHA
- Panel 2: State of the Practice in Identifying and Quantifying Extreme Flood Hazards
- Panel 3: Extreme Precipitation Events
- Panel 4: Flood-Induced Dam and Levee Failures
- Panel 5: Tsunami Flooding
- Panel 6: Riverine Flooding
- Panel 7: Extreme Storm Surges for Coastal Areas
- Panel 8: Combined Events Flooding

A concluding session, Panel 9, provided an opportunity for the Panel co-chairs to summarize their session's presentations and panel discussions. Discussions based on these summaries identified suggested areas for further work which is documented in the Panel 9 report in Chapter 10.

Each panel session is documented as a chapter in these proceedings. The chapters begin with a detailed summary followed by abstracts by the presenters and, in some cases, the panelists. References are provided in the abstracts and panel summaries. Appendix G provides a listing of biographies of the workshop co-chairs, presenters, and panelists.

In total, over 250 participants registered for the workshop, including those viewing remotely using the NRC's webstreaming service. The participants represented Federal employees, contractors, industry representatives, and academics, and countries such as France and the Netherlands were also represented. A list of all workshop attendees and their affiliations can be found in Appendix B.

1.5 Information Sources

To aid the interested reader who wishes to view all of the workshop presentations, please go to the NRC Public Website for meeting archives: http://www.nrc.gov/public-involve/public-meetings/meeting-archives/research-wkshps.html. Appendix E provides a listing of electronic information sources that includes the home Website for the participating Federal agencies with identified special URL sites dealing with extreme precipitation, flooding, dam safety, flood risk management, risk assessments, risk-informed decisionmaking and related published reports and data sources.

PANEL 1

FEDERAL AGENCIES' INTERESTS
AND NEEDS IN PFHA

Co-Chairs:
Nilesh Chokshi, NRC/NRO, and
Mark Blackburn, DOE

Technical Reporters:
Christopher Cook and Maria Pohida, NRC

2. Federal Agencies' Interests and Needs in PFHA

Nilesh Chokshi[1], Mark Blackburn[2], Wendy Reed[3], Christopher Cook[1], and Marie Pohida[1]

[1]Office of New Reactors, U.S. NRC, Rockville, MD
[2]Office of Nuclear Facility Safety Programs, U.S. Department of Energy, Germantown, MD
[3]Office of Nuclear Regulatory Research, U.S. NRC, Rockville, MD

2.1 Motivation

The aim of Panel 1 was primarily to establish the extent to which differing Federal agencies use probabilistic methods in their flood hazard assessment, and to establish the needs of each agency with regards to information required to successfully implement or improve the use of PFHA in their hazard assessments. The presentations included NRC staff's perspectives on the development of a PFHA approach within a risk context. Other presentations focused on probabilistic approaches presently used or under development by the participating agencies, as well as ongoing efforts to develop consensus standards.

2.2 Background

Several Federal agencies have a need for reliable flood hazard data, whether in regards to the potential effects on the safety of a nuclear power plant (NRC), or the potential failure of a dam that the agency owns (USACE, BoR) or regulates (FERC). Traditionally, a deterministic approach to flooding safety has been used; for example, NRC requires that a licensee determine the probable maximum flood (PMF) at a particular site[1]. There is now a trend among Federal agencies to move towards more probabilistic, risk-informed methods.

2.3 Overview of Presentations

Commissioner George Apostolakis was the plenary speaker, providing an overview of the importance of risk-informed, performance-based regulation. He opened his talk by stating that the management of uncertainty with regards to accidents is always a concern. Commissioner Apostolakis provided a history of the NRC's policy on risk-informed, performance-based regulation, which was published as a white paper in 1999[2]. His talk described some of the fundamentals of risk-informed regulation: the risk definition or triplet (What can go wrong? How likely is it? What are the consequences?); issues that are important; deliberative process in making regulatory decisions; and, most importantly, his observation that risk-informed regulation should incorporate insights from both deterministic and risk perspectives, because the issue of incompleteness is common to both perspectives.

Dr. Fernando Ferrante provided further details of the risk-informed framework being used by the NRC, including decision criteria are being applied. He stressed the importance of PFHA to the NRC. He then discussed two components of a probabilistic risk assessment: hazard

---

[1]   NUREG-0800, "Standard Review Plan for the Review of Safety Analysis Reports for Nuclear Power Plants: LWR Edition," Section 2.4.2, Agencywide Documents Access and Management System (ADAMS) Accession No. ML070100647.

[2]   SECY-98-144, " White Paper on Risk-Informed and Performance-Based Regulation," ADAMS Accession No. ML003753601.

assessment and conditional probability of failure assessment, and highlighted the NRC needs in this area.

Dr. Kammerer discussed the Senior Seismic Hazard Analysis Committee (SSHAC) process that is being used in Probabilistic Seismic Hazard Assessment (PSHA), and its central concepts. The SSHAC process has become an increasingly important tool for aiding design and decision making at all levels in both the private sector and government. It was developed in the 1980's because it was observed that data was being legitimately interpreted quite differently by different experts, and these differences of interpretation translated into uncertainties in the numerical results from a PSHA[3]. Dr. Kammerer discussed the applicability of the SSHAC approach to other natural hazards and highlighted the similarity between assessment of seismic and tsunami hazards in terms of basic steps and type of issues that need to be addressed. She provided an overview of how the SSHAC process is carried out; the structured workshops that are held to identify data needs, present expert viewpoints, and examine preliminary models. She highlighted a new NRC report (NUREG-2117[4]) that has been published, which provides an updated procedure on how to implement the SSHAC process.

Dr. England provided BoR perspective and practices related to dam safety. BoR is responsible for the safety of many large dams, most in the western 17 states, and includes some of the highest hazard dams in the US. He highlighted how risk-informed approaches are already an integral part of the BoR dam safety assessments and decision-making, and that there is a lot of overlap with the NRC approach. He also discussed the inter-relationship between deterministic and risk considerations. BoR uses a team based approach (similar to the SSHAC process in that regard) to look at event tree nodes, which are constructed for risk assessment purposes. BoR uses risk in its safety assessments because it allows the BoR to spend resources on the most significant issues.

FERC on the other hand, is only just starting to look at incorporating risk into its regulations. FERC currently regulates 3000 dams, 834 of which are classed as high hazard. David Lord stated that, in 2009, FERC announced in its strategic plan that it was to incorporate risk-informed decision making (RIDM) into its dam safety program. Currently, FERC is working on its RIDM Engineering Guidelines (to include a Hydrologic Hazard Analysis (HHA) chapter, which will use best practices from the BoR). Drafts of these guidelines are to be completed by September 2013.

Discussions by Dr. John Stevenson and Mr. Ray Schneider regarding the activities related to two important flood related ANS standards is very important when it comes to implementation of probabilistic approaches. ANS 2.31, "Determining Design Basis On-site Flooding Caused by Precipitation at Nuclear Facility Sites" is to be made applicable to all types of nuclear facility and sites, even "dry sites" where facility grade is well above any external flood level. The basic concepts used in the standard include probabilistically-defined on-site precipitation and a mean return period in years or mean probability of exceedence/year. The standard will use NOAA Atlas 14 data to extend precipitation data to longer periods. ANS 2.8, "Determining External Flood Hazards for all Nuclear Facilities," was issued in 1992 and withdrawn in 2002. The revision will address climate change considerations and will reflect the changes in prediction tools and probabilistic methods that have occurred over the past 20 years. Some of the features in the standard include a deterministic screening process, a probabilistic method of establishing frequency-based hazard requirements and understanding uncertainties in the

---

[3] NUREG/CR-6372, "Recommendations for Probabilistic Seismic Hazard Analysis: Guidance on Uncertainty and Use of Experts"

[4] NUREG-2117, "Practical Implementation Guidelines for SSHAC Level 3 and 4 Hazard Studies"

hazard. The scope of the standard will include tsunamis and ocean landslides, and dam operation, which were not addressed in the original.

2.4 Summary of Panel Discussion

The following questions were created to guide the discussion period:

1. What are the roles of deterministic and probabilistic hazard analysis in determining a design basis and conducting a risk assessment? How should they complement each other?

2. What is the status of PFHA? For which flood causing mechanisms can PFHA be conducted? What improvements are needed for their use in a risk assessment?

3. Given the inherent large uncertainties, how should these be addressed?

4. What are the impediments, if any, for other flood causing mechanisms to develop PFHA approaches? How can they be overcome?

5. What are your perceptions about the utility and usefulness of a PFHA for your agency missions?

6. Is a formal expert elicitation approach like SSHAC a viable approach for PFHA? What PFHA-specific considerations should be applied?

7. Given the use of PFHA in the development of Design Basis Flooding determination, what is, or should be, the role of Beyond Design Basis Flooding in design and, if required how should it be determined?

Charles Ader, Director of Safety Systems and Risk Assessment in NRC's Office of New Reactors (NRO) opened the comment period. Referring to slide number 8 of Dr. Ferrante's presentation (please see workshop presentations at: http://www.nrc.gov/public-involve/public-meetings/meeting-archives/research-wkshps.html), which details the PRA reactor framework which involves the calculation of core damage frequency (CDF), he stated that new reactors have very low calculated internal event CDF's. The regulations for new reactors require PRAs for all new reactors. Currently, external floods are screened from PRA analysis if the applicant meets deterministic criterion. Dr. Ader stated that he supports flood hazards evaluations, consistent with Commissioner Apostolakis' risk-informed framework. He posed the question: do we have a good perception of risk of facilities if floods are not looked at? (For example, looking at the problem from a systems level, what is the probability of the flood door failing; can a flood door be closed in time for a fast moving event.) He believes that there is a (PRA) role to support design limits. What happens if those design limits are exceeded; is there a cliff effect?

Pat Regan addressed the question of impediments to PFHA (question 4). He fully supports the SSHAC process and thinks it a good idea to come up with a method (to assess uncertainties) similar to SSHAC; FERC are actually using this type of process in some projects. He questioned whether all of the experts needed for a SSHAC process are available at this point and would like to encourage bright young people to grow into flood hazard development. Pat Regan's role at FERC is to develop their risk-informed program. He considers what BoR is doing as being state of the practice in terms of dam safety.

John England of BoR made his comments from a dam safety perspective. Regarding question #1, he stated that BoR uses both deterministic and probabilistic methods on occasion. He stated that BoR was not eliminating maximum concepts, but was using deterministic information complemented with a PRA approach. Regarding question #5, he answered simply that it is required by the BoR.

Fernando Ferrante, regarding question #3, stated that there are large uncertainties. From a risk analyst's perspective, questions associated with flooding include "What is the frequency of PMP and PMF?" Some models (external flooding modes) have wide uncertainties. Need to consider how uncertainties should be characterized and dealt with – even if they are wide. The Integrated Assessments to be performed in response to Fukushima is a deterministic approach which introduces PRA concepts for mitigation. Dr. Ferrante stated that he wants to move away from the idea that wide flood hazard uncertainties are a reason that PFHA can't be done. He also stated that dam failures are also important to the NRC and the agency is trying to understand how these affect plants. He also highlighted the need for timely information: Reactor Oversight Issues are in the public eye and answers are often needed in months.

Annie Kammerer stated that though probabilistic analysis was an imperative for the NRC, deterministic methods are still useful. For example, when a new fault was found around Diablo Canyon that needed an immediate regulatory decision, a deterministic analysis of the plant operability provided a bridge to the SSHAC level 3 process which is still ongoing. The benefits of SSHAC are transparency and detail. The biggest challenge is non-stationary processes such as landslide-induced tsunamis and the high level of sediment transport. USGS (Woods Hole) and NRC found that the chance of a landslide tsunami is influenced by sea level and particulate glaciations, because there is a lower sea level and higher amount of sediment transport and deposition. NRC is currently evaluating seismic risks at the 1E-4 yr today using logic trees.

David Lord made the comment that large uncertainties in the 100-500-year flood make decision making uncomfortable, but that the SSHAC process would force uncertainties to be dealt with.

Commissioner Apostolakis asked the question "What are the top 2 or 3 impediments to having a PFHA completed?" John England's response was that "Folks do not want to try". He elaborated that there are hurdles in education and training; there is still a barrier to technology transfer in the universities. Later, he also added that communication is a challenge and that a multi-disciplinary approach is needed. There is also a notion of a site being absolutely safe because it is designed for PMF versus PRA.

Pat Regan added that the dam safety community is only just moving towards probabilistic: BoR, then USACE, now FERC. In his opinion, the Teton Failure drove the BoR to get into PFHA and the USACE got into PFHA after the New Orleans flooding (Hurricane Katrina). Also, the flooding community was comfortable with deterministic. He stated that a multi-disciplinary approach (geologists, hydrologists, meteorologists) is needed to estimate hazards and many people are reluctant to do this.

Jery Stedinger added that while he was on the National Research Council's 1985 dam safety panel, he found that people fell into 2 camps: 1. Deterministic, PMP, PMF; and 2. Probabilistic. People are more comfortable with PMF. However, there is still a chance that PMF could be exceeded because it is only an estimate!

Commissioner Apostolakis followed up his question with a query about technical issues in implementing PFHA. John England clarified that there are many technical issues.

Geoff Bonnin, NOAA/NWS made a statement regarding uncertainties. He said that at the very extremes, we don't have enough data to know what the distributions are, and to identify the types of events. This lack of knowledge raises the question about applying statistical techniques when we don't know the answer to these things.

2.5 Observations and Insights

- Risk-informed approaches are being used and are being incorporated in safety assessments and decision-making.
- It is not a question of deterministic vs. risk assessment. These are complementary processes.
- A SSHAC-type of approach is viable and has been used to systematically address the issue of uncertainties considering the state-of-knowledge, including lack of data for extreme events.
- What are the impediments to a PFHA? – Willingness to try, availability of experts, communication. (Technical challenges are being met).
- Need for multi-disciplinary teams for assessments and need for incorporating risk analysis in educational systems
- Establish understanding of commonality and differences in risk-informed approaches and decision criteria among the federal agencies.
- Collaborative and coordinated efforts with other federal agencies, industry, standard bodies, and other stakeholders.
- Consider implementation of SSHAC type of approaches for selected hazards.

2.6 Abstracts

The workshop organizing committee developed the flooding topics to be discussed, and chose the panel co-chairs for each panel topic. The co-chairs then identified potential speakers and discussed them with the organizing committee. Following agreement, the co-chairs sent out invitations to the presenters requesting presentation titles and abstracts for documentation in the workshop program. The speakers and panelists are identified in the workshop agenda (please see Appendix A). The following seven abstracts document these presentations, and in some cases, reflect the discussions during the panel session.

2.6.1 U. S. Nuclear Regulatory Commission Staff Needs in Probabilistic Flood Hazard Assessment

Fernando Ferrante, Ph.D.

U.S. Nuclear Regulatory Commission, Washington DC, USA

The U.S. Nuclear Regulatory Commission (NRC) is an independent Federal agency whose mission is to license and regulate the Nation's civilian use of byproduct, source, and special nuclear materials to ensure the adequate protection of public health and safety, promote the common defense and security, and protect the environment [1]. To support its mission, the USNRC performs licensing, rulemaking, incidence response, and oversight activities to ensure the safe operation of critical infrastructure, such as large operating commercial nuclear reactors, research and test reactors, and nuclear fuel cycle facilities.

Specific parts in Title 10 of the *Code of Federal Regulations* [2], which prescribe the requirements under the authority of the USNRC, also include consideration of natural phenomena such as floods. For example, for commercial nuclear reactors, principal design criteria establishing design requirements for structures, systems, and components that provide reasonable assurance that the facility can withstand the effects of natural phenomena without undue risk to the health and safety of the public are described [3]:

(1) Consideration of the most severe of the natural phenomena that have been historically reported for the site and surrounding area, with sufficient margin for the limited accuracy, quantity, and period of time in which the historical data have been accumulated

(2) appropriate combinations of the effects of normal and accident conditions with the effects of the natural phenomena

(3) the importance of the safety functions to be performed

Significant regulatory guidance and research has been developed in order to address the evaluation of various flooding phenomena and mechanisms such as extreme meteorological and oceanographic events (e.g., severe storms, tides, and waves), seiches and tsunamis, and dam failures [references 4 through 9]. Understanding the risks posed to the facilities regulated by the USNRC is also important in terms of resource allocation (e.g., prioritization of inspections and incident response) and emergency planning. It is recognized that, for different natural phenomena, the maturity of available methodologies and data for assessing the likelihood of occurrence of hazards that may challenge plant safety varies significantly from hazard to hazard and can involve wide uncertainty in both the intensity and frequency of the event. The USNRC has spent significant resources to develop and incorporate the use of risk-assessment tools for risk-informed decisionmaking in a variety of applications related to regulatory licensing and oversight [10]. The drivers prompting the staff to use these tools include a policy statement on the use of probabilistic risk assessment (PRA) in regulatory activities [11], as well as guidance for specific activities (e.g., [12]) in which quantitative risk assessments may be used in risk-informed regulatory actions.

In the area of oversight, the USNRC oversees licensees' safety performance through inspections, investigations, enforcement, and performance assessment activities, in which risk tools and risk criteria are routinely applied. In particular, the USNRC has established the Reactor Oversight Process (ROP), which is the Agency's program to inspect, measure, and assess the safety performance of commercial nuclear power plants and to respond to any decline in performance [13]. This includes determining the safety significance of inspection findings through a number of risk-informed activities, including the use of PRA models for individual reactors that are developed and maintained by Idaho National Laboratory (INL) for use by the USNRC [14]. Typically, these models consider a number of potential events that may challenge plant safety, such as loss of AC power necessary to operate critical equipment and/or the loss of capability to cool the nuclear reactor core. As the characteristics of events that may challenge plant safety are identified, the capacity of the safety systems designed to protect critical functions is evaluated for the conditional probability of failure, thus resulting in an overall measure of the available protection with respect to the likelihood of the intervening event. One of the commonly used risk measures in PRA models is the core-damage frequency (CDF), a measure of the likelihood of severe damage to the nuclear fuel used for heat generation in U.S. commercial nuclear reactors.

In many cases, the models are based on events occurring internal to the plant (e.g., random equipment failure, operator errors) while less focus has historically been placed on detailed modeling of extreme natural phenomena, although some studies evaluating the risk contribution of flooding events exist [references 15 through 21]. Typical CDF results for overall plant risk obtained in NRC risk analysis range from an annualized core-damage frequency of 1E-4 per year to 1E-6 per year for all contributing risk scenarios. In the hydrologic community, it has long been recognized that estimating the annualized frequency of severe floods tends to be restrained to a great extent by the historical record available, with significant effort made in the development of methods and approaches to extend frequency estimates beyond typically observed events [22]. The result is that there exists significant uncertainty associated with the ranges of interest of NRC risk applications such as the ones described above, where severe flood estimates may range from 1E-3 per year to 1E-6 per year and below [23}, depending on the quantity and quality of data available as well as the refinement of the methodology used to derive such estimates.

The NRC had been engaged in risk assessment of natural hazards, such as floods, well before the 9.0-magnitude earthquake that struck Japan on March 11, 2011, that eventually led to extensive damage to the Fukushima Dai-ichi nuclear site. However, the formation of the Near-Term Task Force (NTTF) to review insights from this event [24] and the subsequent activities [references 25 through 27] that have included a reevaluation of potential flooding hazards for nuclear facilities regulated by the USNRC have refocused the review and potential enhancement of the treatment and evaluation of events with very low probability but very high consequences with respect to critical infrastructure.

Coupled with the already existing risk framework and tools used in licensing and oversight for commercial nuclear reactors as well as other applications regulated by the NRC, a probabilistic flood hazard assessment (PFHA) effort could provide significant input in the area of flood hazard characterization and the probabilistic treatment of flood protection and mitigation strategies with respect to PRAs. In particular, there is a strong interest for further development in the following areas in applications related to the risk assessment of nuclear facilities licensed and regulated by the NRC:

- Development of methods to consistently estimate annualized flood frequencies in the ranges of interest of USNRC applications with respect to potential contributors to CDF, including extrapolations beyond the available historical record.

- Characterization of a broader spectrum of flood risk contributors, rather than focusing on deterministically derived "worst-case" scenarios, in order to capture the impact of extreme events as well as less severe but more frequent floods.

- Consideration of the feasibility of screening methodologies that take into account the range of locations of facilities licensed and regulated by the USNRC (i.e., with distinct hydrological, meteorological, and geographical characteristics) to identify sites for which certain flooding mechanisms may be more applicable and therefore may require more detailed risk assessment considerations.

- Risk assessment of dam failures (including downstream dam failures that may affect the availability of water from a large body of water relied on to cool various thermal loads during plant operations) on safe operations of nuclear facilities. This would include probabilistic treatment of various individual failure mechanisms, such as overtopping, internal erosion, spillway and hydro-turbine failures, seismic events, and credible combination of events; within the frequency range of interest to NRC applications.

- Possible probabilistic treatment of flood protection structures and barriers (including temporary barriers), while considering potential for degradation from debris impact, erosion, and other effects during severe flooding events.

- Development of a probabilistic assessment of the capacity of mechanical, structural, and electrical systems relied on for safe operation of nuclear facilities to withstand flooding impacts, similar to fragility curves typically associated with seismic risk assessments (e.g., conditional probability of failure with respect to a specific loading or flood level).

- Feasibility of operator manual actions during extreme flooding events in a probabilistic framework, which may be associated with actions such as the installation of flooding protection (e.g., floodgates), construction of barriers (e.g., sandbag barriers), and other actions.

References

[1] U.S. Nuclear Regulatory Commission, "Information Digest 2012–2013," NUREG-1350, Volume 24, August 2012, Agencywide Documents Access and Management System (ADAMS) Accession No. ML12230A174.

[2] *U.S. Code of Federal Regulations*, Chapter I, Title 10, "Energy."

[3] *U.S. Code of Federal Regulations*, "Design bases for protection against natural phenomena," General Criterion 2, in "General Design Criteria for Nuclear Power Plants," Appendix A, to "Domestic Licensing of Production and Utilization Facilities," Part 50, Chapter I, Title 10, "Energy."

[4] U.S. Nuclear Regulatory Commission, "Standard Review Plan for the Review of Safety Analysis Reports for Nuclear Power Plants: LWR Edition," NUREG-0800, Section 2.4.2, "Floods," Rev. 4, March 2007, ADAMS Accession No. ML070100647.

[5] U.S. Nuclear Regulatory Commission, "Standard Review Plan for the Review of Safety Analysis Reports for Nuclear Power Plants: LWR Edition," NUREG-0800, Section 2.4.3: "Probable Maximum Flood (PMF) on Streams and Rivers," Rev. 4, March 2007, ADAMS Accession No. ML070730417.

[6] U.S. Nuclear Regulatory Commission, "Standard Review Plan for the Review of Safety Analysis Reports for Nuclear Power Plants: LWR Edition," NUREG-0800, Section 2.4.4, "Potential Dam Failures," Rev. 3, March 2007, ADAMS Accession No. ML070730417.

[7] U.S. Nuclear Regulatory Commission, "Standard Review Plan for the Review of Safety Analysis Reports for Nuclear Power Plants: LWR Edition," NUREG-0800, Section 2.4.5, "Probable Maximum Surge and Seiche Flooding," Rev. 3, March 2007, ADAMS Accession No. ML070730425.

[8] U.S. Nuclear Regulatory Commission, "Standard Review Plan for the Review of Safety Analysis Reports for Nuclear Power Plants: LWR Edition," NUREG-0800, Section 2.4.6, "Probable Maximum Tsunami Hazards," Rev. 3, March 2007, ADAMS Accession No. ML070160659.

[9] U.S. Nuclear Regulatory Commission, "Design-Basis Flood Estimation for Site Characterization at Nuclear Power Plants in the United States of America," NUREG/CR-7046, November 2011, ADAMS Accession No. ML11321A195.

[10] U.S. Nuclear Regulatory Commission, "A Proposed Risk Management Regulatory Framework - A report to NRC Chairman Gregory B. Jaczko from the Risk Management Task Force," April 2012, ADAMS Accession No. ML12109A277.

[11] U.S. Nuclear Regulatory Commission, "Use of Probabilistic Risk Assessment Methods in Nuclear Regulatory Activities; Final Policy Statement," *Federal Register*, Vol. 60, No. 158, August 16, 1995, pp. 42622–42629.

[12] U.S. Nuclear Regulatory Commission, "An Approach for Using Probabilistic Risk Assessment In Risk-Informed Decisions on Plant-Specific Changes to the Licensing Basis," Regulatory Guide 1.174, July 1998, ADAMS Accession No. ML003740133.

[13] U.S. Nuclear Regulatory Commission, "Reactor Oversight Process," NUREG-1649, Rev. 4, December 2006, ADAMS Accession No. ML070890365.

[14] U.S. Nuclear Regulatory Commission, "Systems Analysis Programs for Hands-on Integrated Reliability Evaluations (SAPHIRE) Version 8," NUREG/CR-7039, June 2011. http://www.nrc.gov/reading-rm/doc-collections/nuregs/contract/cr7039/

[15] U.S. Nuclear Regulatory Commission, "Perspectives Gained from the Individual Plant Examination of External Events (IPEEE) Program - Volume 1," NUREG-1742, Section 4.1 – end, April 2002. ADAMS Accession No. ML021270122.

[16] U.S. Nuclear Regulatory Commission, "Shutdown Decay Heat Removal Analysis of a Babcock and Wilcox Pressurized Water Reactor - Case Study," NUREG/CR-4713, March 1987.

[17] U.S. Nuclear Regulatory Commission, "Shutdown Decay Heat Removal Analysis of a Westinghouse 2-Loop Pressurized Water Reactor - Case Study," NUREG/CR-4458, March 1987.

[18] U.S. Nuclear Regulatory Commission, "Shutdown Decay Heat Removal Analysis of a General Electric BWR3/Mark I - Case Study," NUREG/CR-4448. March 1987.

[19] U.S. Nuclear Regulatory Commission, "Shutdown Decay Heat Removal Analysis of a Combustion Engineering Pressurized Water Reactor - Case Study," NUREG/CR-4710. March 1987.

[20] U.S. Nuclear Regulatory Commission, "Shutdown Decay Heat Removal Analysis of a Westinghouse 3-Loop Pressurized Water Reactor - Case Study," NUREG/CR-4762. March 1987

[21] U.S. Nuclear Regulatory Commission, "Shutdown Decay Heat Removal Analysis of a General Electric BWR4/Mark I - Case Study," NUREG/CR-4767, March 1987.

[22] U.S. Department of the Interior, Bureau of Reclamation, "Guidelines for Evaluating Hydrologic Hazards," June 2006, Denver, CO, available at http://www.engr.colostate.edu/~pierre/ce_old/classes/ce717/Hydrologic_Hazard_Guidelines_final.pdf.

[23] Jeff Harris et al., "Estimating Probability of Extreme Events," *World Environmental and Water Resources Congress 2009: Great Rivers*, American Society of Civil Engineers, 2009, pp. 1–9, available at http://dx.doi.org/10.1061/41036(342)624.

[24] U.S. Nuclear Regulatory Commission, "Recommendations for Enhancing Reactor Safety in the 21st Century - The Near-Term Task Force Review of Insights from the Fukushima Dai-Ichi Accident," July 2011, ADAMS Accession No. ML111861807.

[25] U.S. Nuclear Regulatory Commission, "Proposed Orders and Requests for Information in Response to Lessons Learned from Japan's March 11, 2011, Great Tohoku Earthquake And Tsunami," Commission Paper SECY-12-0025, February 17, 2012. ADAMS Accession No. ML12039A103.

[26] U.S. Nuclear Regulatory Commission, "Request for Information Pursuant to Title 10 of the *Code of Federal Regulations* 50.54(f) regarding Recommendations 2.1, 2.3, and 9.3, of the Near-Term Task Force Review of Insights from the Fukushima Dai-ichi Accident," March 12, 2012, ADAMS Accession No. ML12053A340.

[27] U.S. Nuclear Regulatory Commission, "Guidance for Performing the Integrated Assessment for Flooding - DRAFT Interim Staff Guidance," JLD-ISG-2012-05, Revision 0 (Draft Issue for Public Comment), September 20, 2012, ADAMS Accession No. ML12235A319.

2.6.2 Probabilistic Hazard Assessment Approaches: Transferable Methods from Seismic Hazard

Annie M. Kammerer

U.S. NRC, Office of New Reactors/Office of Nuclear Regulatory Research, Rockville, MD 20852

Quantitative assessment of all natural hazards shares common attributes and challenges. These include the need to determine the best estimate and uncertainty of the hazard levels, limited data, high levels of uncertainty, and the need to include expert judgment in the assessment process. Over the last several decades, approaches for assessing seismic hazard have matured and current approaches are transferable to other types of hazard. While the level of uncertainty in seismic hazard remains high, particularly in low- to moderate-seismicity regions, highly structured methods of expert interaction and model development have been developed and applied with success. In 1997, the Senior Seismic Hazard Analysis Committee (SSHAC) developed a structured, multilevel assessment framework and process (the "SSHAC process"), described in NUREG/CR-6372, "Recommendations for Probabilistic Seismic Hazard Analysis: Guidance on Uncertainty and Use of Experts," that has since been used for numerous natural hazard studies since its publication. In 2012, the NRC published NUREG-2117, "Practical Implementation Guidelines for SSHAC Level 3 and 4 Hazard Studies," after studying the experience and knowledge gained in the application of the original guidance over the last 15 years. NUREG-2117 provides more detailed guidelines that are consistent with the original framework described in NUREG/CR-6372, which is more general and high level in nature. NUREG-2117 provides an extensive discussion of the Level 3 process, which is well suited to the assessment of other natural hazards.

When seismic hazard assessments are conducted for critical facilities such as nuclear power plants, the judgments of multiple experts are required to capture the complete distribution of technically defensible interpretations (TDIs) of the available earth science data. The SSHAC process provides a transparent method of structured expert interaction entirely focused on capturing the center, body, and range (CBR) of the full suite of TDIs of the available data. The goal is not to determine the single best interpretation; rather, it is to develop and integrate all TDIs into a logic-tree framework wherein the weights of the various branches are consistent with the level to which the data and information available supports the interpretation. This approach leads to greater assurance that the "true" hazard at a site is captured within the breadth of the probabilistic seismic hazard assessment (PSHA) results.

To achieve this, a study prepared using the SSHAC process goes through a series of steps that can be separated into evaluation and integration phases. The fundamental goal of the SSHAC Level 3 process is to properly carry out and completely document the activities of evaluation and integration, defined as follows:

Evaluation: The consideration of the complete set of data, models, and methods proposed by the larger technical community that are relevant to the hazard analysis.

Integration: Representing the CBR of TDIs in light of the evaluation process (i.e., informed by the assessment of existing data, models, and methods).

As discussed in detail in NUREG-2117, the process includes a number of well-defined roles for participants, as well as three required workshops, each with a specific objective. The evaluation process starts with the Technical Integrator (TI) team identifying (with input from resource and proponent experts) the available body of hazard-relevant data, models, and methods—including all those previously produced by the technical community—to the extent possible. This body of existing knowledge is supplemented by new data gathered for the study. The first workshop is focused on the identification of hazard-relevant data, models, and methods. The TI team then evaluates these data, models, and methods and documents both the process by which this evaluation was undertaken and the technical bases for all decisions made regarding the quality and usefulness of these data, models, and methods. This evaluation process explicitly includes interaction with, and among, members of the technical community. The expert interaction includes subjecting data, models, and methods to technical challenge and defense. The second workshop provides a forum for proponents of alternative viewpoints to debate the merits of their models. The successful execution of the evaluation is confirmed by the concurrence of the Participatory Peer Review Panel (PPRP) that the TI team has provided adequate technical bases for its conclusions about the quality and usefulness of the data, models, and methods, and has adhered to the SSHAC assessment process. The PPRP will also provide guidance on meeting the objective of considering all of the views and models existing in the technical community.

Informed by this evaluation process, the TI team then performs an integration process that may include incorporating existing models and methods, developing new methods, and building new models. The objective of this integration process is to capture the CBR of TDIs of the available data, models, and methods. The technical bases for the weights on different models in the final distribution, as well as the exclusion of any models and methods proposed by the technical community, need to be justified in the documentation. The third workshop provides an opportunity for the experts to review hazard-related feedback on their preliminary models and to receive comments on their models from the PPRP. To conclude the project satisfactorily, the PPRP will also need to confirm that the SSHAC assessment process was adhered to throughout and that all technical assessments were sufficiently justified and documented.

References

U.S. Nuclear Regulatory Commission, "Recommendations for Probabilistic Seismic Hazard Analysis: Guidance on Uncertainty and Use of Experts," NUREG/CR-6372, April 1997, Agencywide Documents Access and Management System (ADAMS) Accession Nos. ML080090003 and ML080090004.

U.S. Nuclear Regulatory Commission, "A Performance-Based Approach to Define the Site-Specific Earthquake Ground Motion," Regulatory Guide 1.208, March 2007, ADAMS Accession No. ML070310619.

U.S. Nuclear Regulatory Commission, "Practical Implementation Guidelines for SSHAC Level 3 and 4 Hazard Studies," NUREG-2117, Rev. 1, April 2012, ADAMS Accession No. ML12118A445.

2.6.3 Bureau of Reclamation Dam Safety Probabilistic Flood Hazard Analysis Perspective

John F. England, Jr.[1] and Karen M. Weghorst[1]

[1]U.S. Department of Interior, Bureau of Reclamation, Flood Hydrology and Consequences Group, 86-68250, Technical Service Center, Denver, CO 80225

The Bureau of Reclamation uses risk analysis to assess the safety of dams, recommend safety improvements, and prioritize expenditures (Reclamation, 2011a). Reclamation has been using risk analysis as the primary support for dam safety decisionmaking for over 15 years, and has developed procedures to analyze risks for a multitude of potential failure modes. By definition, "risk" is the product of the likelihood of an adverse outcome and the consequences of that outcome. The likelihood of an adverse outcome is the product of the likelihood of the loading that could produce that outcome and the likelihood that the adverse outcome would result from that loading. Risk analysis, from a hydrologic perspective, requires an evaluation of a full range of hydrologic loading conditions and possible dam failure mechanisms tied to consequences of a failure.

The Bureau of Reclamation conducts risk analysis at different levels, from screening level analyses performed by an individual (with peer review) during a Comprehensive Facility Review (CFR), to full-blown facilitated team risk analyses, which include participation by "field" personnel. One key input to a risk analysis is the loading characteristic; this is generally in the form of hydrologic hazard curves or seismic hazard curves for floods and earthquakes. Manuals, guidelines, and practical reference material detailing risk-analysis methodology for dam safety applications based on Reclamation's experience are available (Reclamation, 2011b). This presentation provides an overview of the Dam Safety Act, Reclamation's risk process, and risk reduction. Risk analysis at the Bureau of Reclamation has evolved over the years and will continue to evolve.

References

Bureau of Reclamation (1999), A Framework for Characterizing Extreme Floods for Dam Safety Risk Assessment. Prepared by Utah State University and Bureau of Reclamation, Denver, CO, November.

Bureau of Reclamation (2002), Interim Guidelines for Addressing the Risk of Extreme Hydrologic Events. Bureau of Reclamation, Denver, CO, August, 3 p. (England, J.F., principal author).

Bureau of Reclamation (Reclamation) (2011a), Dam Safety Public Protection Guidelines, A Risk Framework to Support Dam Safety Decision-Making, Bureau of Reclamation, Denver, Colorado, 31 p. http://www.usbr.gov/ssle/damsafety/documents/PPG201108.pdf

Other reports related to Reclamation's Public Protection Guidelines are here: http://www.usbr.gov/ssle/damsafety/references.html

Bureau of Reclamation (Reclamation) (2011b) Dam Safety Risk Analysis Best Practices Training Manual. Bureau of Reclamation, in cooperation with US Army Corps of Engineers, Denver, CO, Aug., version 2.2, http://www.usbr.gov/ssle/damsafety/Risk/methodology.html

2.6.4 FERC Need for Probabilistic Flood Hazard Analysis (PFHA) Journey From Deterministic to Probabilistic

David W. Lord, P.E.

Senior Civil Engineer, Risk-Informed Decision-Making, FERC, Division of Dam Safety and Inspections, Portland Regional Office

The Federal Energy Regulatory Commission (FERC) has a well-developed program (Ref. 1) to calculate a deterministic value for the probable maximum flood (PMF) based on the probable maximum precipitation (PMP). However, problems with these calculations increasingly cause us concern. The PMF determinations use the precipitation estimate from the applicable Hydro Meteorological Reports (HMR) or increasingly use a site-specific or statewide study (Note a) which updates the general concepts in the HMRs and applies them with some rigor to a smaller area. The application of these updated PMP studies have taken out much of the conservatism inherent in the HMR estimates of PMP because they have reduced the PMP estimates up to 56 percent (Note b).

Another problem is that the PMP and, thus PMF numbers are treated as having zero probabilities, i.e., as upper bounds that can't be exceeded, rather than having a relatively low probability. For some parts of the country, farther from the warm ocean moisture source, these estimates are likely to have very low annual exceedance probabilities, about 1×10^{-7} or lower. However, in coastal regions closer to the warm moisture source, the estimates of the PMP (Ref. 2) and thus PMF probability may be much higher, possibly several orders of magnitude higher, possibly between 1×10^{-3} and 1×10^{-5}. This results in both uneven treatment between dams, and the possibility of flood risks that have not been adequately evaluated because the PMP and resulting PMF is a relatively frequent event.

Another fundamental issue is that many dam overtopping failure modes are from a combination of circumstances, such as trash buildup on spillway gates or inability to open a spillway gate during a relatively frequent flood. The frequency of events between 1×10^{-2} and 1×10^{-4} might be of greater concern in combination with these events than an extreme flood event.

The FERC's modern dam safety program started with implementing regulations in 1981. The FERC's Division of Dam Safety and Inspections (D2SI) has regulatory responsibility for approximately 3000 dams and approximately 950 dams classified as having high and significant hazard potential. The "high hazard" classification is based on whether generally worst-case dam failures would have the potential to cause loss of life. "Significant hazard" dams are reserved for economic or sensitive environmental issues. As discussed above, our dam safety program has relied on generally conservative deterministic analyses until recently.

In 2000, D2SI established a Potential Failure Modes Analysis (PFMA) program for all high and significant hazard (HSH) dams. In 2009, the Commission adopted a strategic plan to develop a probabilistically oriented program, titled Risk-Informed Decision Making (RIDM). D2SI recently completed a Screening Level Portfolio Risk Analysis for all HSH dams, and has initiated development of risk-informed engineering guidelines included a Hydrologic Hazard Analysis (HHA) chapter. Drafts of these guidelines are to be completed by September 2013.

Reclamation has a well-respected program for developing estimates for the annual exceedance probability (AEP) of extreme precipitation and flood events and for developing resulting reservoir exceedance curves. Our new HHA chapter will rely heavily on the current Best Practices used by Reclamation. However, we are a regulator and not a dam owner like Reclamation, so our guidelines will require us to be able to communicate with hundreds of dam owners and dozens of consultants.

While Reclamation can make an informed decision based on the information they develop, we will have to be able to use information primarily developed by the owners and consultants in consultation with our staff. Some of the questions about extreme flood frequency are as follows:

- Should the PMP be the upper bound of any frequency estimates of extreme storms?

- Are there simple procedures for rare to extreme flood and precipitation frequency estimates to judge whether more robust estimates are needed at a dam?

- What circumstance (i.e., level of risk) would require an owner to develop new flood frequency estimates at a dam (i.e., when does a probabilistic estimate augment or replace a deterministic estimate?)?

- If the current PMP and PMF estimates are inadequate at a dam, do we complete new deterministic estimates, or should we rely only on new probabilistic calculations?

Notes

(a) Statewide PMP studies have been approved for Wisconsin and Michigan in 1993 and Nebraska in 2008. Statewide PMP studies are ongoing for Arizona, Ohio, and Wyoming.

(b) Reduced the PMP value of the HMR for a 24-hour, 20,000-mi^2 storm from 6.8 inches to 3.0 inches in western Nebraska.

References

(1) Federal Energy Regulatory Commission, "Engineering Guidelines for the Evaluation of Hydropower Projects," Chapter 8, "Determination of the Probable Maximum Flood," September 2001, available at
http://www.ferc.gov/industries/hydropower/safety/guidelines/eng-guide/chap8.pdf.

(2) Schaefer, M.G., "PMP and Other Extreme Storms: Concepts and Probabilities," *Dam Safety '95: 1995 ASDSO Annual Conference Proceedings*, Association of State Dam Safety Officials, Lexington KY, October 1995.

2.6.5 Development of On-Site Precipitation in a Probabilistic Format

John Stevenson, Consultant Engineer, ANS 2.8 Working Group

6611 Rockside Rd., Suite 110, Independence, Ohio 44131

Introduction

There are potentially a large number of natural phenomenon hazards (NPH) that can result in the unmitigated release of radioactive material and waste from failure of structures, systems and components (SSCs) in nuclear facilities that are important to nuclear safety as listed in Table 2–1. The potential for the natural hazard phenomenon to be a design basis for important nuclear-safety structures, systems and components in the nuclear facility is a function of the five criteria listed in Table 2–2.

The NPH that have generally been identified as design bases for the important-to-safety SSCs and for which NRC, DOE, and ANS siting and design standards have been developed are:

- earthquake (Refs. 1, 2)
- tornado, hurricane, and straight winds (Refs. 3, 4, 5)
- flooding external to the site (Ref. 6)
- onsite precipitation (Ref. 7)

It should be noted that the stated safety goals for both NRC and DOE nuclear facilities are probabilistic in nature and, in general, require the natural hazards to be defined in terms of a mean probability of exceedance or mean return periods, which currently is the case for earthquake and wind hazards phenomena. At this time, the NRC defines the external site flooding and onsite precipitation phenomena deterministically based on the concepts of probable maximum flood (PMF) and probable maximum precipitation (PMP). Hence, their contribution to the probabilistic safety goal cannot be rationally determined. Mr. David Lord (FERC) mentioned in his presentation during the Panel 1 session that the estimated equivalent mean probability of exceedance for the PMF and PMP was somewhere between 10^{-3} and 10^{-7} per year depending on the geographic region of interest in the United States. The published DOE Standard 1020-2012 (Ref. 8) and ANS Draft Standard 2.8 for flooding external to the site and the ANS 2.31 Draft Standard for onsite precipitation hazards have already been or are in the process of being modified or developed to define offsite flooding and onsite precipitation on a probabilistic basis in order to be compatible with NRC and DOE safety goals.

The status of the ANS 2.31 Standard, currently under development and described at the PFHA workshop, is a national consensus standard intended to define onsite precipitation phenomena on a probabilistic basis which recognizes the need to define precipitation phenomena using a graded approach. This requires different return periods in design as a function of the severity of the potential for unmitigated radiological releases from the facility to the environment.

The ANS siting standards for natural phenomena hazards use a graded approach that is currently divided into five natural hazard design categories as a function of their unmitigated level of radiation release as typically identified in the ANS 2.26-2004 Standard. This categorization with some revision is recognized in the DOE Standard 1020-2012. As shown in Table 2–3, it is hoped that in the future, design of nuclear facilities compliant with 10 CFR

Parts 50, 52, 70 and 72, that a similar probabilistically defined, graded approach will be considered for use.

Table 2–1 Natural Hazards That Might Affect Nuclear Facilities' Radiological Safety and the Criteria Typically Used to Determine Whether These Hazards Will . Become Design-Basis Events

Natural Phenomenon Hazards	See Table 2–2
1. Earthquake	0
2. Straight Wind and Hurricanes and Their Missiles	0.5
3. Tornado Wind, Differential Pressure, and Missiles	0
4. Precipitation – Snow Onsite	0
5. Precipitation – Rain Onsite	0
6. Lightning	0
7. Volcanoes	3, 4
8. Electromagnetic Pulses	1
9. Avalanche	3, 4
10. Biological	1
11. Climate Change	1
12. Drought	1
13. Flood External to the Site	0
14. Fog	2
15. Forest Fire	1, 4
16. Frost	2
17. Hail	2
18. High Summer Temperature	2
19. Ice Cover	2*
20. Landslide	3, 4
21. Low Lake or River Water Level	2
22. Low Winter Temperature	2
23. Damaging Meteorite	3
24. Sandstorm	1, 4
25. Seiche	0.5
26. Soil Shrink Swell Consolidation	1
27. Storm Surge	0.5
28. Tsunami	0.5
29. Waterspout	4
30. Wave Action	0.5

*Except for external distribution systems.

Table 2–2 Criteria Typically Used To Evaluate Design Basis for Nuclear Facilities

0.	A phenomenon generally considered as a design basis for a nuclear facility.
1.	A phenomenon which occurs slowly or with adequate warning with respect to the time required to take appropriate protective or remedial action.
2.	A phenomenon which in itself has no significant effect on the safe or shutdown operation of a nuclear facility.
3.	An individual phenomenon which by itself has a mean probability of occurrence generally less than 10^{-5} per year or a mean return period of less than 100,000 years.
4.	The nuclear facility is located a sufficient distance from or above the postulated NPH so as to eliminate or mitigate its affects to an acceptable level.
5.	A phenomenon which is included or enveloped by design for another phenomenon. For example, storm-surge and Seiche are included in external to the site flooding; or straight wind or hurricane effects are enveloped by the tornado phenomenon.

Table 2–3   Guidance for Natural Phenomenon Hazard (Seismic) Design Category (SDC) Based on Unmitigated Consequences of SSC Failures

Category	Unmitigated Consequence of SSC Failure	
	Worker	Public
SDC-1[1]	No radiological or chemical release consequences but failure of SSCs may place facility workers at risk of physical injury.	No consequences
SDC-2	Lesser radiological or chemical exposures to workers than those in SDC-3 below in this column as well as placing more workers at risk. This corresponds to the criterion in Table 2–1 that workers will experience no permanent health effects.	Lesser radiological and chemical exposures to the public than those in SDC-3 below in this column, supporting that there are essentially no offsite consequences as stated in Table 2–1.
SDC-3	0.25 Sv (25 rem) < dose < 1 Sv (100 rem) AEGL[2]-2, ERPG-2 < concentration < AEGL-3, ERPG[3]-3. Concentrations may place emergency facility operations at risk, or place several hundred workers at risk.	0.05 Sv (5 rem) < dose < 0.25 Sv (25 rem) AEGL-2, ERPG-2 < concentration < AEGL-3, ERPG-3
SDC-4	1 Sv (100 rem) < dose < 5 Sv (500 rem) concentration > AEGL-3, ERPG-3	0.25 Sv (25 rem) < dose < 1 Sv (100 rem) > 300 mg soluble uranium intake, concentration > AEGL-3, ERPG-3
SDC-5	Radiological or toxicological effects may be likely to cause loss of facility worker life.	1 Sv (100 rem) < dose, concentration > AEGL-3, ERPG-3.

[1]   "No radiological or chemical release consequences" or "No consequences" means that material releases that cause health or environment concerns are not expected to occur from failures of SSCs assigned to this category.
[2]   Acute Exposure Guideline Levels
[3]   Emergency Response Planning Guidelines

Status of ANS 2.31

PowerPoint Presentation – "Status of the Draft ANS 2.31 Standard, Determining Design Basis On-Site Flooding Caused by Precipitation at Nuclear Facility Sites."
http://pbadupws.nrc.gov/docs/ML1305/ML13057A717.pdf

References

1. American National Standards Institute and American Nuclear Society, "Categorization of Nuclear Facility Structures, Systems, and Components for Seismic Design," ANSI/ANS-2.26-2004 (reaffirmed 2010), La Grange Park, IL.

2. American Society of Civil Engineers and Structural Engineering Institute, "Seismic Design Criteria for Structures, Systems, and Components in Nuclear Facilities," ASCE/SEI 43-05, Reston, VA.

3. American National Standards Institute and American Nuclear Society, "Standard for Estimating Tornado, Hurricane, and Extreme Straight Line Wind Characteristics at Nuclear Facility Sites," ANSI/ANS-2.3-2011, La Grange Park, IL.

4. U.S. Nuclear Regulatory Commission, "Design Basis Tornado for Nuclear Power Plants," Regulatory Guide 1.76, Agencywide Documents Access and Management System (ADAMS) Accession No. ML070360253.

5. U.S. Nuclear Regulatory Commission, "Design-Basis Hurricane and Hurricane Missiles for Nuclear Power Plants," Regulatory Guide 1.221, ADAMS Accession No. ML110940300.

6. American National Standards Institute and American Nuclear Society, "Determining Design Basis Flooding at Power Reactor Sites," ANSI/ANS-2.8-1992 (withdrawn 2002), La Grange Park, IL.

7. U.S. Nuclear Regulatory Commission, "Design Basis Floods for Nuclear Power Plants," Regulatory Guide 1.59, ADAMS Accession No. ML003740388.

8. U.S. Department of Energy, "Natural Phenomenon Hazard Analysis and Design Criteria for DOE Facilities," Standard 1020-2012, Washington, DC.

2.6.6 ANSI/ANS-2.8 Determining External Flood Hazards for Nuclear Facilities; Workgroup Status

Dr. Yan Gao
Presented by Mr. Ray Schneider

ANSI/ANS-2.8, "Determining External Flood Hazards for Nuclear Facilities," is under development. This short presentation will focus on the current status and planned activities of the Working Group (WG) for this year.

Historically, ANSI/ANS-2.8 (1992), "Determining design basis flooding at Nuclear Power sites," was administratively withdrawn in 2002.

The current ANSI/ANS-2.8 being developed is to fill an important nuclear standards gap. The new standard will reflect nuclear-site flooding events since 1992, insights from Hurricane Katrina surges, record Mississippi and Red River floods, and combined flooding events from European coastal sites. To more appropriately address the external flood risks, the revised standard will focus on a probabilistic process for assigning the flood hazard. In addition, other upgrades to the standard will include consideration of surge effects of tsunamis, integrated consideration of "climate change" effects, state-of-the-art enhancements in technology, computation methodologies and capabilities in fluid dynamics and hydrology.

The scope of the standard includes riverine flooding (rainfall, snowmelt, controlled and uncontrolled releases from dams), dam failure (hydrologic and non-hydrologic caused by seismic, intrinsic, and other phenomena), hurricane-induced storm surge, wind- and earthquake-generated seiche and tsunami (initiated seismically and by landslide). The standard does not include conditions associated with low water; dispersion, dilution; and travel time of accidentally released effluents; ground water; or internal and external flooding from pipe or tank failures.

Using a probabilistic approach, the standard will provide graded requirements for the construction and use of probabilistically generated external flood hazards, including requirements for treatment of aleatory and epistemic uncertainty. The resulting hazards could then be represented as illustrated in Figure 1 as a family of curves of hazard frequencies of exceedance vs. flood elevations.

Currently, the Working Group has 21 members representing various stakeholders in the industry.

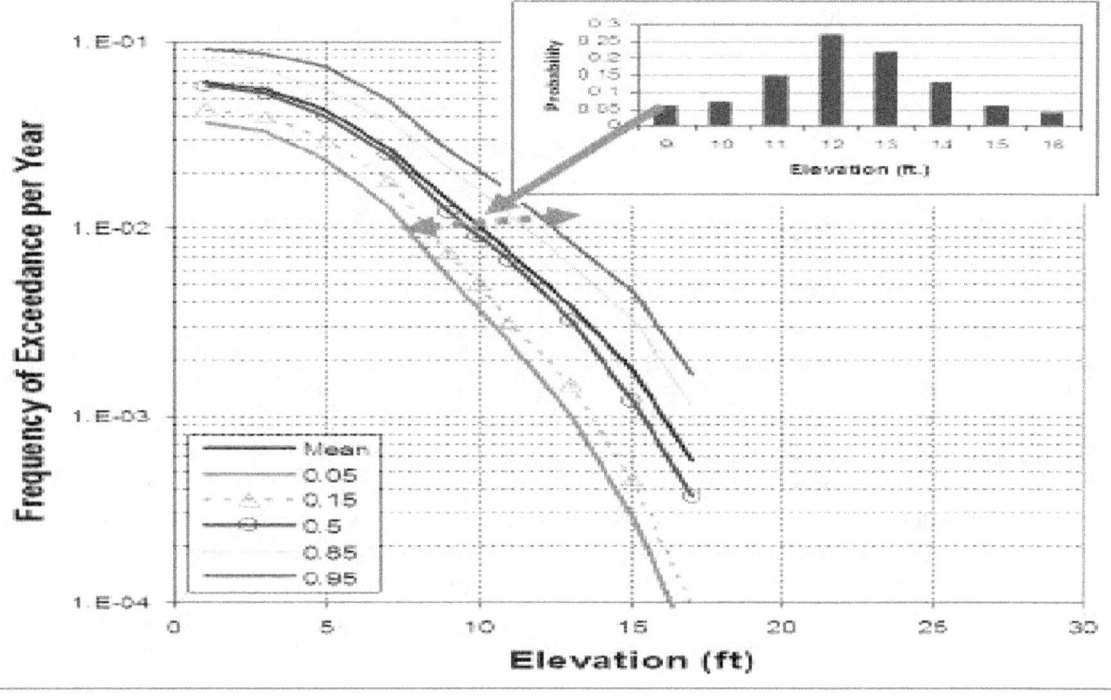

Figure 2–1 Curves of hazard frequencies of exceedance vs. flood elevations

Presentation at: http://pbadupws.nrc.gov/docs/ML1305/ML13057A718.pdf

2.6.7   U.S. Department of Energy and Probabilistic Flood Hazard Assessment

W. Mark Blackburn, P.E.[1]

[1]Director, Office of Nuclear Facility Safety Programs, HS-32 Germantown (GTN) DOE

The U.S. Department of Energy (DOE) has used Probabilistic Flood Hazard Analysis for many years. The usage of Probabilistic Flood Hazard Analysis is endorsed in a recently approved DOE document called DOE Standard DOE-STD-1020-2012, *Natural Phenomena Hazards Analysis and Design Criteria for Department of Energy Facilities*. This document references a 1988 paper entitled *Preliminary Flood Hazard Estimates for Screening Models for Department of Energy Sites* by McCann and Boissonnade. This current endorsement has its roots in two archived documents: 1) DOE Standard 1022-94, *Natural Phenomena Hazards Characterization Criteria*; and 2) DOE Order 6430.1A, *General Design Criteria*. All of these documents recommended usage of Probabilistic Flood Hazard Analysis for sites which had flood as one of the predominant design basis events. Because this technique is used at our sites, we are interested in learning about its usage in other agencies which may have similar applications as those in DOE.

PANEL 2

STATE OF THE PRACTICE IN IDENTIFYING AND QUANTIFYING EXTREME FLOOD HAZARDS

Co-Chairs:
Timothy Cohn, USGS
Wilbert Thomas, Michael Baker, Jr., Inc.

Technical Reporters:
Joseph Giacinto, Mark McBride, and Randy Fedors, NRC

3. State of the Practice in Identifying and Quantifying Extreme Flood Hazards

Timothy Cohn[1], Wilbert Thomas[2], Joseph Giacinto[3], Mark McBride[3] and Randy Fedors[4]

[1]U.S. Geological Survey, Reston, VA
[2]Michael Baker, Jr. Inc., Manassas, VA
[3]Office of New Reactors, U.S. NRC, Rockville, MD
[4]Office of Nuclear Material Safety and Safeguards, U.S. NRC, Rockville, MD

3.1 Motivation

Extremely large floods can cause the greatest damage, but they are also the rarest flood events. Because they are rare, information on them is sparse, so estimating their magnitude and frequency of occurrence is difficult. However, because the largest floods are associated with the greatest risk of flood damage, estimating their frequency is important for planning civil works, land use, preventive measures, and emergency operations.

3.2 Background

The estimation of flood hazards has traditionally been based primarily on instrumental records of floods from the last 100 years or less. This short historical record provides only sparse information about extremely large floods, and is unlikely to include many of the largest floods that could occur at the site under present conditions. Recent efforts, developments, and applications over the past 10 years on extreme flood hazards have focused on including historical and paleoflood data that span several thousand years. New methods, based on additional kinds of data and improved statistical methods, are likely to improve capabilities for estimating the frequency and magnitude of extreme floods and the resulting risks associated with them.

Panel 2, "State of the Practice in Identifying and Quantifying Extreme Flood Hazards," addressed such improved methods. Its overall emphasis was on the use of probabilistic flood hazard assessment (PFHA) methods for quantifying flood risks. This approach quantifies the increasing magnitude of floods with increasing recurrence interval. Among practitioners, it is increasingly preferred over methods, such as the "probable maximum flood," that use a single value to characterize flood risks.

3.3 Overview of Presentations

Wilbert O. Thomas of Michael Baker, Jr., Inc. gave a historical survey of flood frequency analysis in the United States. His emphasis was on the development of statistical methods for estimating flood frequency at gauged sites with observed data on annual peak flows. This process began with publication of the first such method in 1914. Improvements and standardization have continued to the present, much of this driven by Federal agencies and working groups.

Jery Stedinger of Cornell University discussed the basic concepts, philosophy, and strategy underlying current analytical methods for extreme floods. He described current efforts to update published guidance using additional observations made over the last 30 years. He emphasized new methods, including better use of historical information from informal observations and paleofloods to extend the flood record, regionalizing multiple flood records to improve extrapolation to extreme events, and addressing uncertainty in flood analyses (including the additional uncertainty posed by climate change).

Jim O'Connor of the U.S. Geological Survey described methods for inferring the timing, magnitude, and frequency of floods that occurred before historical observations and presented a case study from the Black Hills, SD, as an example. These inferences are based on physical evidence; for example, sediments deposited in slackwater environments above the level of historically observed floods. Sediment deposits may be dated by methods such as radiocarbon analysis and optically stimulated luminescence. Discharges are inferred by hydraulic analysis based on estimated maximum flood stage.

Douglas J. Clemetson described methods currently used by the U.S. Army Corps of Engineers for quantifying the hazards of extreme floods. Six different methods are used to estimate extreme flood probabilities, ranging from well-established extrapolation of frequency curves to innovative use of paleoflood information. Because there is no good way to develop confidence intervals for frequencies, sensitivity analyses are performed to examine the range of probabilities.

John England described new hydrologic hazard methods developed and applied over the past 15 years by the U.S. Bureau of Reclamation to assess dam safety risk. He emphasized the need for having a variety of methods available for use, with the selection of any one method depending on the objectives and data available, or the selection of multiple methods to provide confidence in results. Rather than using single upper bounds for discharge and reservoir level, the Bureau develops hydrologic hazard curves that relate peak flow and volume to annual exceedance probability. These are based on a number of statistical approaches, broadly classified as streamflow-based or rainfall-and-runoff-based. Results are combined and weighted by a team of hydrologists and are used as the basis for case-specific risk analysis.

3.4   Summary of Panel Discussion

The panel members discussed questions proposed by the moderator and audience:

What has kept the Federal government from adopting probabilistic flood hazard assessment across agencies?

One reason is the engineering culture within Federal agencies. The continuity of this culture slows progress. There is still much use of setpoints, such as the probable maximum flood. Although these appear to be exact, they create a false sense of confidence, such that these values are taken as being final in some evaluations. What is needed among the Federal agencies is a dedicated, focused, and intense effort to characterize the process and methods for PFHA. Improving expertise within agencies will be slow and gradual, although this can be speeded by cooperation and joint efforts among agencies.

Another problem is lack of consistency of practice across agencies. For example, many definitions are being applied to characterize uncertainty. Practitioners are often not clear what is meant when they define "uncertainty" in their estimates.

Probably the largest problem is deficient expertise in PFHA. There are a number of areas of deficiency. Academic engineering and earth science programs generally do not require the courses in probability and statistics needed to apply PFHA methods to flood frequency analysis. Multiple courses in statistics are needed, at levels beyond the introductory. Currently, in-house statistical training is necessary because new employees cannot be expected to have the necessary background.

Expertise is also lacking in paleoflood analysis. Paleoflood research is inherently hard; the average engineer does not have the necessary background. Paleoflood investigations do not lend themselves to "cookbook" approaches because of the variety of conditions near different rivers, and because of the variety of skills needed. One person can't have all the needed skills. A team is needed, including such skills as geology, hydraulics, surveying, and sediment dating. The Bureau of Reclamation has had good results over the last 15 years from using such teams.

How far back in time can the record of paleofloods be used?

From 1,000 to 5,000 years is typical for the western United States. The best records are from deposits protected in caves and alcoves, but flood plain deposits can be usable for several thousand years. We use records only from the present climate regime. Records from more than 10,000 years ago, which reflect the last ice age when climate conditions were fundamentally different, are not useful for projections of the historical record.

How is the stratigraphic record translated to flood stage and to discharge? Was sediment transport considered? How far upstream and downstream does the estimate go?

The basic question is how to translate the paleoflood record to flood discharge. Different discharges are routed through the reach of interest using one- or two-dimensional hydraulic models, based on assumptions such as that the channel has not changed and that the flood behaved as water flow rather than debris flow. Different discharges are tried until the calculated flood stage is consistent with the elevations of paleoflood sediment deposits. Sediment transport is very important, but is also very complicated, and was not incorporated in the Black Hills study. The Bureau of Reclamation practice is to address channel hydraulics, but not sediment transport. For some purposes, the volume of a flood is more important than peak discharge. Methods for estimating volumes of paleofloods are poorly developed.

The uncertainty that results from making indirect discharge computations is not limited to paleofloods. Many large floods are not gaged by the USGS, but discharges are instead computed by indirect methods based on observed flood levels and channel hydraulic characteristics.

Uncertainty analysis is conducted to determine the effects of various types of uncertainty. For example, uncertainty in discharge measurements for very large floods can be small compared to the large variability among such floods. Uncertainty analysis must be conducted, but it may show that measurement uncertainty is not important.

What are the contributions of climate data and climate change to the extreme flood estimates, and how does climate change impact estimates?

The estimates use many kinds of climate data, in many forms. One of their main purposes is to reduce uncertainty in the estimates. Incorporation of climate change impacts can take many forms, including integration of recent point data with regional or areal historical data. The effects are hard to estimate, but clearly are highly variable and location-dependent, with some areas becoming drier and others wetter. It is not clear whether any suitable methods are now available. This is a topic for future research.

Does the "cookbook" style of analysis presented in Bulletin 17B inhibit innovation, particularly since different agencies and institutions have different needs?

The applicability of Bulletin 17B was intended to be limited. This bulletin was designed for applications such as levee and floodplain management, and was not intended for extending estimates to 1-in-10,000 events and for identifying outliers. Agencies understand that more complex methods are needed for extreme flood events. Use of Bulletin 17B is not mandatory and was not intended to limit agencies. It was intended to provide a uniform approach across agencies, partly to avoid litigation that could result if different Federal agencies produced different estimates. Also, the "cookbook" methods may have the beneficial effect of stimulating university research to address their deficiencies.

NOAA's Hydrometeorological Reports (e.g., HMR-41) give the impression that only 72-hour durations of precipitation are of concern. Dr. England described use of longer antecedent periods (up to about 14 days) for an evaluation of the Trinity River dam. What are the criteria for using a shorter or longer duration of precipitation event than is discussed in the HMRs?

Investigators are increasingly emphasizing pre-storm conditions, and sequences of related hydrologic events, rather than just examining a single fixed period. For example, an evaluation of the May 2010 flood in Nashville includes examining two sets of factors: meteorological conditions (such as antecedent precipitation over the entire month of May, and the sequence of storms) and flood conditions (such as reservoir levels and soil moisture). For another example, in the Southeast a sequence of back-to-back hurricanes might have to be considered.

The point is, rather than using a fixed duration, we should look at sequencing of events from the recent record, and then possibly expand the period of antecedent conditions. We need to decide what sequence of events we want to model. This may require extending the period to pull in significant outlier events.

For the Trinity River Study, the critical conditions were the reservoir itself and the sequence of storms that caused streamflow to fluctuate. We wanted to bracket the whole sequence of significant events, so we used the flexibility of being able to look at a longer period.

How are orographic effects handled?

The U.S. Army Corps of Engineers develops areal reduction factors based on historic storms or else based on the HMRs. The Bureau of Reclamation uses a hybrid approach, using NOAA Atlas 14, "Precipitation-Frequency Atlas of the United States" (where applicable), with customized extensions. The approach is a mix of point-frequency and storm-based analysis, depending on the situation. Orographic effects, and how to address boundary effects, are examples of concerns that arise when applying regionalization methods to extend records to better define extreme events.

3.5   Observations and Insights

- Academic programs in engineering and earth science should encourage advanced statistical training relevant to analysis of extreme floods, and programs providing training in paleoflood analysis should be expanded.

- Federal agencies should standardize nomenclature used to discuss risk and uncertainty.

- Federal agencies should make a dedicated, focused, and intense effort to characterize the process and methods for PFHA.

- Federal agencies should provide more institutional support for multidisciplinary teams for complex analyses, such as of paleofloods.

- Hydrologists involved in estimating extreme flood hazards should pay close attention to ongoing research on the effects of climate change, and should incorporate insights from this research into their estimates.

- Hydrologists involved in estimating extreme flood hazards should make careful analyses of pre-storm conditions over relevant periods, rather than using only fixed event durations.

- Research is needed on how to incorporate sediment transport into indirect estimates of discharge.

- Research is needed on how to estimate paleoflood volumes in addition to peak discharges.

- Research is needed on improving methods for characterizing uncertainty in estimates of extreme flood hazards.

3.6 Abstracts

The workshop organizing committee developed the flooding topics to be discussed and chose the panel co-chairs for each panel topic. The co-chairs then identified potential speakers and discussed them with the organizing committee. Following agreement, the co-chairs sent out invitations to the presenters requesting presentation titles and abstracts for documentation in the workshop program. The speakers and panelists are identified in the workshop agenda (please see Appendix A). The following three abstracts document these presentations and, in some cases, reflect the discussions during the panel session.

3.6.1 Overview and History of Flood Frequency in the United States

Wilbert O. Thomas, Jr.

Michael Baker, Jr., Inc., Manassas, VA 20110

Characterizing the frequency of flood peak discharges is essential in designing critical infrastructure such as dams, levees, bridges, culverts, and nuclear power plants and in the definition of flood plains for land -use planning. Estimates of flood frequency are needed for gaged and ungaged locations but this discussion is focused primarily on frequency analysis on gaged streams with observed flood data. Statistical flood frequency analysis provides the risk of structures being overtopped or flood plain areas being inundated.

Statistical flood frequency analysis in the United States started with a paper by Fuller (1914) who analyzed long records of daily flows and peak flows around the world, but particularly from the United States. Fuller (1914) developed equations for estimating floods of a given return period T as a function of drainage area and is typically credited with having the first published formula or equation involving flood frequency (Dawdy et al., 2012). In his 1914 paper, Fuller also discussed plotting positions for analyzing flood distributions and suggested plotting at the median which later became known as the Hazen plotting position. Fuller (1914) apparently used plotting positions for the at-site frequency analysis and then developed equations for estimating the T-year flood discharges for ungaged locations.

Hazen (1914), in his discussion of Fuller's paper, presented for the first time the concept of probability paper and argued for the use of the lognormal distribution as a model of peak flows (Dawdy et al., 2012). Hazen (1930) describes his plotting position method in detail in his 1930 book "Flood Flows" and provides examples of estimating flood discharges for New England streams. Hazen's approach involved plotting flood data on graph paper with the normal probability scale on the horizontal axis (percent chance exceedance) and the ratio to the mean annual flood on a logarithmic vertical axis (lognormal probability paper) and adjusting for the skew in the data. The Hazen plotting position estimates the exceedance probability as (2m-1)/2N where m is the order number (starting with one as the largest) and N is the total number of data points. This plotting position assigns a return period of 2N for the largest flood in N years of record.

The use of plotting positions was the prevalent approach for flood frequency analysis in the period from 1930 to 1967. Beard (1943) developed the median plotting position (m-0.30/N+0.4 with m and N as defined above) and this plotting position is still being used by the U.S. Army Corps of Engineers (USACE) for graphical display of flood data. In the 1940s, the U.S. Geological Survey (USGS) adopted the Weibull plotting position (m/N+1) as described by Kimball (1946) for estimating T-year flood discharges (Dalrymple, 1960; Benson, 1962). The USGS approach was to plot flood data on Gumbel probability paper with either a logarithmic or arithmetic vertical scale for discharge and draw a smooth curve through the data. The graphical procedures required significant engineering judgment and there was variation among engineers and hydrologists in applying these methods.

Beard's 1952 and 1962 publications on "Statistical Methods in Hydrology" documented flood frequency procedures being used by USACE in this time period (Beard, 1952, 1962). By 1962 the USACE was using the log-Pearson Type III method that involved fitting the logarithms of the annual peak flows to the Pearson Type III distribution. Foster (1924) introduced the Pearson system of frequency distributions to the American engineering profession and applied these methods to the study of annual runoff for gaging stations in the Adirondack Section of New York. Foster (1924) used the untransformed data but Beard (1962) recommended the use of logarithms of the annual peak data.

In April 1966, the Subcommittee on Hydrology under the Inter-Agency Committee on Water Resources (IACWR) published Bulletin No. 13 on "Methods of Flow Frequency Analysis" (IACWR, 1966). Five methods of flow frequency analysis **currently in common use** were described: Hazen, Pearson Type III, Gumbel, gamma, and graphical distribution-free methods. This appears to be the first interagency effort to describe methods for flood frequency analysis. Bulletin 13 summarized and described several methods of determining flood frequency but did not recommend any particular methods. No testing or comparison of the various methods was attempted.

In August 1966, the 89th Congress passed House Document No. 465 entitled "A Unified National Program for Managing Flood Losses." This document recommended the establishment of a panel of the Water Resources Council (WRC) to "present a set of techniques for frequency analyses that are based on the best known hydrological and statistical procedures". House Document No. 465 also recommended the creation of a national flood insurance program that was eventually created in 1968. In response to House Document No. 465, the Executive Director of WRC in September 1966 assigned the responsibility for developing a uniform technique to the WRC Hydrology Committee. The Hydrology Committee established a Work Group on Flow-Frequency Methods comprised of members of 12 Federal agencies and two statistical consultants. The work group applied the following six methods at 10 long-term stations (records greater than 40 years):

- 2-parameter gamma distribution
- Gumbel distribution
- log-Gumbel distribution
- log-normal distribution
- log-Pearson Type III distribution
- Hazen method

In December 1967, the work group published Bulletin No. 15, "A Uniform Technique for Determining Flood Flow Frequencies," with the recommendation to fit the logarithms of the annual peak discharges to a Pearson Type III distribution using the method of moments (USWRC, 1967). The reasons for adopting this method were:

1. Some Federal agencies were already using the log-Pearson Type III distribution (e.g., USACE) and computer programs were available.

2. The log-normal is a special case of both the Hazen and log-Pearson Type III method with zero skew.

3. The log-Pearson Type III distribution uses skew as third parameter and it was felt this would provide more flexibility and applicability in estimating flood discharges nationwide.

4. The Hazen method was partially graphical and required more subjectivity in its application.

Manuel Benson, USGS, who chaired the Bulletin 15 work group published a journal article that provided additional details on the analyses performed by the work group (Benson, 1968). It soon became evident that Bulletin 15 was not as uniform a method as originally conceived. Some of the reasons were non-uniform treatment of outliers, skew, and historical data. In January 1972, the Hydrology Committee of WRC initiated review of Bulletin 15 and the need for more uniform guidelines. The Work Group on Flood Flow Frequency of the Hydrology Committee was comprised of members of 12 Federal agencies. In March 1976, the Hydrology Committee of WRC published Bulletin 17, "Guidelines for Determining Flood Flow Frequency". Bulletin 17 recommended the continued use of the log-Pearson Type III distribution with the method of moments for parameter estimation (USWRC, 1976). This recommendation was based on a study that Beard (1974), who was at the University of Texas, conducted for the work group; the study applied eight combinations of distributions and estimation techniques to annual peak discharges at 300 long-term gaging stations (with records longer than 30 years). To correct problems noted with Bulletin 15, Bulletin 17 recommended the use of a low outlier test for censoring low peaks, use of generalized (or regional) skew based on a study by Hardison (1974), and a mathematical procedure for adjusting for historical data based on a weighted-moments procedure suggested by Fred Bertle of the Bureau of Reclamation.

Shortly after Bulletin 17 was published, it was noted that there was a discrepancy about the order of the historical adjustment and the determination of weighted skew. In June 1977, Bulletin 17A, "Guidelines for Determining Flood Flow Frequency," was published; it clarified that the historical adjustment was to be applied before the weighting of skew (USWRC, 1977). This clarification is the only significant difference between Bulletins 17 and 17A. A few editorial corrections were also made.

With time, problems with the Bulletin 17A methodology began to surface. These problems can be summarized as follows:

(1)   The low-outlier test did not adequately identify all the low outliers.

(2)   There was some confusion over the estimation and use of generalized skew.

(3)   There were inconsistencies in the use of the conditional probability adjustment for low outliers.

In September 1981, the Hydrology Committee of WRC published Bulletin 17B, "Guidelines For Determining Flood Flow Frequency" (USWRC, 1981). The Work Group on Revision of Bulletin 17 of the Hydrology Committee was comprised of members of six Federal agencies: the Soil Conservation Service, USACE, USGS, National Weather Service, Bureau of Reclamation, and the Tennessee Valley Authority. The following technical changes were in Bulletin 17B to correct problems noted in Bulletin 17A:

(1)   Revised guidelines for estimating and using generalized skew

(2)   A new procedure for weighting generalized and station skew based on research by Tasker (1978) and Wallis et al. (1974)

(3)   A new test for detecting high outliers and a revised test for detecting low outliers based on research by Grubbs and Beck (1972)

(4)   Revised guidelines for the application of the conditional probability adjustment
In March 1982 several editorial corrections were made in Bulletin 17B and the guidelines were republished by the Hydrology Subcommittee of the Interagency Advisory Committee on Water Data (IACWD). The new sponsorship of Bulletin 17B was necessitated by the dissolution of

the Water Resources Council in the fall of 1982. The Hydrology Subcommittee membership of the six Federal agencies remained the same under the two different sponsors. Bulletin 17B, "Guidelines for Determining Flood Flow Frequency," published in March 1982 by the Hydrology Subcommittee of IACWD, is still the set of guidelines used by Federal agencies and most state and local agencies for flood frequency analyses for gaged streams (IACWD, 1982).

In January 2000, a Hydrologic Frequency Analysis Work Group (HFAWG) was formed to consider improvements in the current guidelines for hydrologic frequency analysis computations (e.g. Bulletin 17B) and to evaluate other procedures for frequency analysis of hydrologic phenomena. The HFAWG is a work group under the Hydrology Subcommittee of the Advisory Committee on Water Information (ACWI). The HFAWG, the Hydrology Subcommittee, and the ACWI are all comprised of Federal and non-Federal organizations. This is a change from the past where Bulletins 15, 17, 17A and 17B were developed solely by Federal agencies.

The HFAWG has met several times since January 2000 and the minutes of those meetings are on the HFAWG web site at http://acwi.gov/hydrology/Frequency/. In November 2005 the HFAWG developed a plan for updating Bulletin 17B (see web site for details). The **major** improvements in Bulletin 17B that are being evaluated include:

- Evaluating and comparing the performance of a new statistical procedure, the Expected Moments Algorithm (EMA), to the weighted-moments approach of Bulletin 17B for analyzing data sets with historical information;

- Evaluating and comparing the performance of EMA to the conditional probability adjustment of Bulletin 17B for analyzing data sets with low outliers and zero flows;

- Describing improved procedures for estimating generalized (regional) skew; and

- Describing improved procedures for defining confidence limits.

During 2012, considerable progress has been made in testing and evaluating the EMA procedure and in developing a new test for detecting influential low peaks. A March 2012 report titled "Updating Bulletin 17B for the 21st Century" that is posted on the HFAWG web site describes the results of testing EMA and current Bulletin 17B techniques. It is anticipated that a revised draft version of Bulletin 17B will be developed in 2013.

In summary, at-site flood frequency techniques have been evolving in the United States over the last 100 years. Since the publication of House Document No. 465 in August 1966, the emphasis has been on developing a uniform and consistent technique that can be used by all Federal agencies. Thomas (1985) and Griffis and Stedinger (2007) describe in more detail the evolution of Bulletin 17 to the current Bulletin 17B technique. Ongoing research and testing by the HFAWG should result in an improved version of Bulletin 17B in 2013.

References

Beard, L.R., 1943, "Statistical analysis in hydrology," *Transactions of the American Society of Civil Engineers* 108:1110–1160.

Beard, L.R., 1952, "Statistical methods in hydrology," Civil Works Engineer Bulletin 52-24, U.S. Army Corps of Engineers, Sacramento, CA.

Beard, L.R., 1962, "Statistical methods in hydrology," Civil Works Investigations Project CW-151, U.S. Army Corps of Engineers, Sacramento, CA, http://water.usgs.gov/osw/bulletin17b/Corps_Beard_1962.pdf.

Beard, L.R., 1974, "Flood Flow Frequency Techniques," CRWR-119, Center for Research in Water Resources, Austin, TX: University of Texas, http://water.usgs.gov/osw/bulletin17b/Beard_FFFT_1974.pdf.

Benson, M.A., 1962, "Evolution of Methods for Evaluating the Occurrence of Floods," Geological Survey Water-Supply Paper 1580-A, U.S. Geological Survey, Washington, DC, http://pubs.usgs.gov/wsp/1580a/report.pdf.

Benson, M.A., 1968, "Uniform Flood-Frequency Estimating Methods for Federal Agencies," *Water Resources Research* 4(5):891–908.

Dalrymple, T., 1960, "Flood-Frequency Analyses, Manual of Hydrology: Part 3. Flood-Flow Techniques," Geological Survey Water-Supply Paper 1543-A, U.S. Geological Survey, Washington, DC, http://pubs.usgs.gov/wsp/1543a/report.pdf.

Dawdy, D.R., V.W. Griffis, and V.K. Gupta, 2012, "Regional Flood-Frequency Analysis: How We Got Here and Where We Are Going," *Journal of Hydrologic Engineering* 17(9):953–959, http://dx.doi.org/10.1061/(ASCE)HE.1943-5584.0000584.

Foster, H.A., 1924, "Theoretical frequency curves and their applications to engineering problems," *Transactions of the American Society of Civil Engineers* 87:142–203.

Fuller, W.E., 1914, "Flood flows," *Transactions of the American Society of Civil Engineers* 77:567–617.

Griffis, V.W., and J.R. Stedinger, 2007, "Evolution of Flood Frequency Analysis with *Bulletin 17*," *Journal of Hydrologic Engineering* 12(3):283–297, http://dx.doi.org/10.1061/(ASCE)1084-0699(2007)12:3(283).

Grubbs, F.E., and G. Beck, 1972, "Extension of Sample Sizes and Percentage Points for Significance Tests of Outlying Observations," *Technometrics* 14(4):847–854.

Hardison, C.H., 1974, "Generalized Skew Coefficients of Annual Floods in the United States and Their Application," *Water Resources Research* 10(5):745–752.

Hazen, A., 1914, "Discussion of 'Flood flows' by W. E. Fuller," *Transactions of the American Society of Civil Engineers* 77:626–632.

Hazen, A., 1930, *Flood Flows: A Study of Frequencies and Magnitudes*, John Wiley & Sons, New York, NY.

Interagency Advisory Committee on Water Data 1982, "Guidelines For Determining Flood Flow Frequency," Bulletin No. 17B of the Hydrology Subcommittee, Office of Water Data Coordination, U.S. Geological Survey, Reston, Virginia, 183 p. http://water.usgs.gov/osw/bulletin17b/dl_flow.pdf.

Inter-Agency Committee on Water Resources (IACWR), 1966. Methods of Flow Frequency Analysis. Bulletin 13 of Subcommittee on Hydrology, 42 p.

Kimball, B.F., 1946, "Assignment of frequencies to a completely ordered set of sample data," *Transactions, American Geophysical Union* 27(6):843–846.

Tasker, G.D., 1978, "Flood Frequency Analysis with a Generalized Skew Coefficient," *Water Resources Research* 14(2):373–376.

Thomas, W.O., Jr., 1985, "A Uniform Technique For Flood Frequency Analysis," *Journal of Water Resources Planning and Management*, ASCE, 111(3):321–337, http://dx.doi.org/10.1061/(ASCE)0733-9496(1985)111:3(321).

U.S. Water Resources Council (USWRC), 1967. "A Uniform Technique for Determining Flood Flow Frequencies," Bulletin No. 15 of the Hydrology Committee, Washington, DC, http://water.usgs.gov/osw/bulletin17b/Bulletin_15_1967.pdf.

USWRC, 1976, "Guidelines for Determining Flood Flow Frequency," Bulletin No. 17 of the Hydrology Committee, Washington, DC.

USWRC, 1977, "Guidelines for Determining Flood Flow Frequency," Bulletin No. 17A of the Hydrology Committee, Washington, DC.

USWRC, 1981, "Guidelines for Determining Flood Flow Frequency," Bulletin No. 17B of the Hydrology Committee, Washington, DC.

Wallis, J.R., N.C. Matalas, and J.R. Slack, 1974, "Just a Moment!", *Water Resources Research* 10(2):211–219.

3.6.2 Quantitative Paleoflood Hydrology

Jim O'Connor[1]

[1]U.S. Geological Survey, Oregon Water Science Center, oconnor@usgs.gov; 503-251-3222

Paleoflood hydrology (Kochel and Baker, 1982) is the reconstruction of the magnitude and frequency of past floods using geological or botanical evidence. The following synopsis of paleoflood hydrology is derived from Benito and O'Connor (in press). Over the last 30 years, paleoflood hydrology has achieved recognition as a new branch of geomorphology and hydrology (Baker, 2008), employing principles of geology, hydrology, and fluid dynamics to infer quantitative and qualitative aspects of unobserved or unmeasured floods on the basis of physical evidence left behind. Flood evidence includes various geologic indicators (flood deposits and geomorphic features) and flotsam deposits, as well as physical effects on vegetation. Resulting inferences can include timing, magnitude, and frequency of individual floods at specific sites or for specific rivers, as well as conclusions regarding the magnitude and frequency of channel-forming floods. The obvious benefit of paleoflood studies is obtaining information on floods from times or locations lacking direct measurements and observations. Findings from paleoflood studies support flood hazard assessments as well as understanding of the linkages between climate, land use, flood frequency, and channel morphology.

Paleoflood studies typically take one of two forms: (1) analyses focused on determining quantitative information for specific events, such as the timing, peak discharge, and maximum stage of an individual flood or floods; and (2) studies investigating more general spatial and temporal patterns of flooding, commonly to assess relations among climate, land use, flood frequency, and magnitude, and geomorphic response (such as channel morphology or floodplain sedimentation and erosion processes). Both types of investigations share approaches and techniques, but, in general, studies of specific paleofloods are most typically conducted in bedrock or otherwise confined river systems for which preservation of stratigraphic and geomorphic records of individual floods are more likely (Kochel and Baker, 1982), although many studies have obtained valuable paleoflood information for alluvial rivers (Knox, 1999; Knox and Daniels, 2002). Studies relating channel form or floodplain morphology to past flood characteristics more typically are conducted for alluvial river corridors, and follow from the classic studies of Schumm (1968) and Dury (1973). In general, quantitative information of specific events is likely to be most appropriate for assessing risk to nuclear facilities; consequently, the emphasis of this section is on studies that can provide specific information on the timing, magnitude, and frequency of individual floods.

Quantitative paleoflood hydrology relies on identification of evidence of flooding in conjunction with application of hydrodynamic principles to determine flow magnitude. These two aspects of investigation typically lead to four phases of analysis: (1) documentation and assessment of flood evidence; (2) determination of paleoflood ages; (3) estimation of flow magnitude, typically peak discharge, associated with flood evidence; and (4) incorporation of paleoflood data into the flood frequency analysis. The first component is geological and archival, requiring historical research, geomorphology, stratigraphy, and sedimentology, and the second involves geochronology in order to identify and date physical evidence of flooding. The third component requires hydraulic analysis to assign a flow magnitude to paleoflood evidence. A common final step is to incorporate paleoflood discharge and chronology information into a flood frequency analysis to determine peak flows associated with probability quantiles. These paleoflood studies generally are more successful and have fewer uncertainties in fluvial systems with resistant boundaries, such as bedrock or semi-alluvial channels. These environments, because of stable depositional sites, tend to have longer and clearer stratigraphic records of floods—sometimes exceeding several thousand years—and have stable boundary conditions, leading to greater confidence in using present topography to determine past hydraulic conditions.

Most paleoflood studies have focused on semiarid and arid regions, although studies have successfully extended flood records in humid environments as well (e.g., Fanok and Wohl, 1997; Springer and Kite, 1997; and Kidson and others, 2005). In general, paleoflood studies extend flood records by hundreds or thousands of years and commonly provide compelling evidence of flood discharges exceeding those of the observation record (Enzel and others, 1993; O'Connor and others, 1994; Hosman and others, 2003; and Harden and others, 2011). In certain areas, very large floods cluster on time scales of decades and centuries, interpreted as a response to climate variability (Ely and others, 1993; Knox 2000; and Benito and others, 2003a).

Slackwater Flood Records

Paleoflood records are derived from physical evidence of flood stage. The best high-water marks include mud, silt, seed lines, and flotsam (e.g., fine organic debris, grass, and woody debris) that closely mark peak flood stage. This type of evidence typically only persists for weeks, in humid climates, to several years, in semiarid and arid climates. But more lasting evidence can also be preserved, including fine-textured flood sediment (slack-water flood deposits), gravel and boulder bars, silt lines, and erosion features, as well as botanical evidence such as scars on riparian trees. Depending on the environment, such evidence can persist for millennia (Baker, 1987, 2008; Kochel and Baker, 1988; and Webb and Jarrett, 2002).

The most complete paleoflood records generally result from analysis of stratigraphic sequences of fine-grained flood deposits found in slack-water and eddy environments. Slackwater flood deposits are fine-grained sedimentary deposits that accumulate from suspension during floods (Baker and others, 2002). Slackwater sedimentation areas include flooded valley margins subject to eddies, back-flooding, flow separation, and water stagnation during high stages. Diminished flow velocities in these areas promote rapid deposition of the fine-grained fraction of the suspended load. The resulting slack-water flood deposits commonly contain sedimentary structures and textures reflecting flow energy, direction, and velocities.

Slack-water depositional environments can be any location of diminished flow velocity or flow separation, typically (1) areas of channel widening, (2) severe channel bends, (3) obstacle hydraulic shadows where flow separation causes eddies, (4) alcoves and caves in bedrock walls, (5) back-flooded tributary mouths and valleys, or (6) high surfaces flanking the channel. In narrow bedrock reaches, slackwater flood deposits are commonly found in adjacent caves or alcoves or under rock overhangs.

Paleoflood Chronology

Developing a flood chronology is key to assessing flood frequency and commonly requires numerical age dating of sedimentary flood units and intervening deposits. Numerical dating underlies chronologies, typically by radiocarbon and optically simulated luminiscence (OSL). Radiocarbon dating is the most common absolute dating tool employed in paleohydrologic work, although OSL dating is becoming more common. Organic materials such as wood, charcoal, seeds, and leaf fragments are entrained by floods and commonly deposited in conjunction with clastic sediment in slackwater sequences. Additionally, flood deposits may cover vegetation or organic cultural materials, and can in turn be covered by vegetation and organic detritus. All of these types of materials can be radiocarbon dated, thereby providing information on the age of enclosing or bounding flood deposits. Organic materials most likely to provide high-fidelity constraints on flood ages are those not likely to have persisted for a long period of time before deposition, such as seeds, fine organic detritus, and twigs. Commonly, however, radiocarbon dating is performed on charcoal contained within flood deposits, which can persist for hundreds or thousands of years before being incorporated in a flood deposit.

For most studies, it is assumed that radiocarbon ages from detrital material within flood deposits closely approximates the flood date, although the most conservative assumption is that the radiometric date provides a maximum limiting age for the enclosing deposit. This is particularly the case for radiocarbon dates from detrital charcoal. Dating of in-situ organic materials, such as charcoal from ground fires between affected surfaces bracketed by flood deposits or pedogenic carbon between flood deposits, can provide robust constraints on the timing of flood sequences. As for most geologic investigations, dating of multiple organic materials and multiple deposits within a stratigraphical profile increases confidence in flood age determinations. The 5730-year half-life of carbon-14 limits radiocarbon dating to deposits less than 40,000 years old. Also, radiocarbon dating suffers from significant imprecision for the period 1650 to 1950 AD because of the significant fossil fuel burning and introduction of variable amounts of carbon-14 into the atmosphere during the industrial revolution.

The OSL method (Aitken, 1998) is a dating technique which indicates the burial time of deposits, principally quartz and feldspar minerals. This approach allows determination of when sediment was last exposed to light ("bleached"). For the purposes of dating sequences of flood deposits, the general presumption is that the sediment was last exposed to light before deposition. Sampling and analysis involves several steps of collecting and analysis of sand-sized sediment from a target deposit without inadvertent exposure to light. Developments in OSL instrumentation are reducing the sample size to individual quartz and feldspar grains. Moreover, new analytical protocols have improved the application OSL dating for alluvial deposits, resulting in numerical dating with age uncertainties within 5 to 10%, even for g deposits less than 300 years old (Arnold and others, 2009). The technique can be hampered (1) when the proper species of quartz are not present in the deposits and (2) for floods in which the transported sediment was not bleached by exposure to light, either because of high turbidity levels or because the flood occurred at night. But under appropriate conditions, OSL dating can be an important tool, especially for deposits (1) containing little or no organic material, (2) older than the range of radiocarbon dating (>40,000 years), or (3) more recent than 300 years, in which case radiocarbon dating cannot yield precise results.

Radiocarbon and OSL dating can be supplemented with analysis of modern radionuclides such as caesium-137 and lead-210 (Ely and others, 1992). Both of these isotopes were introduced into the atmosphere during nuclear bomb testing in the 1950s, and their presence in flood deposits signifies a post-1950 age. Likewise, human artifacts, such as beer cans (House and Baker, 2001), pottery (Benito and others, 2003a), and other archaeological materials can provide numeric age constraints on enclosing deposits.

Dendrochronology has supported several paleoflood studies because of the identifiable responses of tree growth to damage of the bark and wood-forming tissues, buds, and leaves, and to radial growth following partial uprooting of the trunk (Yanosky and Jarrett, 2002; Jacoby and others, 2008). For situations in which flood damage or effects can be related to tree-ring chronologies derived from the affected tree or from established regional chronologies, flood ages can commonly be determined to the specific year, and in some instances a specific season (Sigafoos, 1964; and Ruíz-Villanueva and others, 2010).

Paleoflood Discharge Estimation

Hydraulic analysis is the basis for discharge estimates for most quantitative paleohydrologic studies. In most analyses, discharge estimates follow from the assumption that the elevation of paleostage evidence provides a minimum estimate of the maximum stage attained by an identified flood. In some situations, deposit elevations may closely approximate the maximum flood stage, although this assumption is difficult to verify (Jarrett and England, 2002) except for specific investigations of height differences between flood indicators and actual flood water depth for modern floods (Kochel 1980, Springer and Kite, 1997, Jarrett and England, 2002; and House and others, 2002). Uncertainties in the fidelity of paleoflood stage evidence to actual maximum stages can be evaluated in conjunction with most discharge estimation procedures. A number of formulae and models are available to estimate flood discharge from known water surface elevations (O'Connor and Webb, 1988; and Webb and Jarrett, 2002), ranging from simple hydraulic equations to more involved, multidimensional hydraulic modeling. Most paleoflood studies assume one-dimensional flow with calculations based on (1) uniform flow equations (e.g., the Manning equation), (2) critical flow conditions, (3) gradually varied flow models, and (4) one-dimensional St. Venant equations. In complex reaches, multidimensional modeling may reduce uncertainties associated with reconstructing flood discharge (Denlinger and others, 2002; and Wohl, 2002). As described by Webb and Jarrett (2002), the appropriate approach for a particular site depends on local hydraulic conditions.

Incorporating Paleoflood Information in Flood Frequency Analysis

Paleoflood information provides tangible information on the occurrence and magnitude of large and infrequent floods. Although paleoflood information may not be as precise as gaged or observed records and is not continuous in the manner of many measurement programs, understanding of the timing and magnitude of the largest floods can substantially reduce uncertainties in flood quantile estimates when considered in a statistically appropriate manner. Several statistical methods have been applied to estimate distribution function parameters for paleoflood datasets. The most efficient methods for incorporating imprecise and categorical data are: (1) maximum likelihood estimators (Stedinger and Cohn, 1986), (2) the method of expected moments (Cohn and others, 1997; and England and others, 2003a) and (3) Bayesian methods (O'Connell and others, 2002; O'Connell, 2005; and Reis and Stedinger, 2005). Some examples of employing these techniques for flood frequency analysis using both gaged and paleoflood records include O'Connor and others (1994), Bureau of Reclamation (2002), England and others (2003b), Levish and others (2003), Hosman and others, (2003), Thorndycraft and others, (2005a), England and others (2010), and Harden and others (2011). In nearly all cases, the addition of paleoflood information greatly improves estimates of low-probability floods, most commonly indicated by markedly narrower confidence limits about flood quantile estimates.

References

Aitken, M.J., 1998, *An Introduction to Optical Dating: The Dating of Quaternary Sediments by the Use of Photon-stimulated Luminescence*, Oxford University Press, Oxford, UK.

Arnold, L.J., et al., 2009, "A revised burial dose estimation procedure for optical dating of young and modern-age sediments," *Quaternary Geochronology* 4:306–325.

Baker, V.R., 1987, "Paleoflood hydrology and extreme flood events," *Journal of Hydrology* 96:79–99.

Baker, V.R., 2008, "Paleoflood hydrology: Origin, progress, prospects," *Geomorphology* 101(1–2):1–13.

Benito, G., et al., 2003a, "Palaeoflood record of the Tagus River (Central Spain) during the Late Pleistocene and Holocene," *Quaternary Science Reviews* 22:1737–1756.

Benito, G., and O'Connor, J.E., in press, "Quantitative paleoflood hydrology," Fluvial Geomorphology Volume of *Treatise in Geomorphology*, E.E. Wohl (ed.), Elsevier.

Bureau of Reclamation, 2002, "Flood Hazard Analysis - Folsom Dam, Central Valley Project, California," Denver, CO, January 2002, ftp://ftp.usbr.gov/jengland/Dam_Safety/Folsom_FloodHazard_Report_all.pdf.

Cohn, T.A., Lane, W.L., Baier, W.G., 1997, "An algorithm for computing moments-based flood quantile estimates when historical flood information is available," *Water Resources Research* 33:2089–2096.

Denlinger, R.P., O'Connell, D.R.H., House, P.K., 2002, "Robust determination of stage and discharge: an example form an extreme flood on the Verde River, Arizona," House, P.K., et al. (eds.), *Ancient Floods, Modern Hazards: Principles and Applications of Paleoflood Hydrology*, Water Science and Application Series, Vol. 5, American Geophysical Union, Washington, DC.

Dury, G.H., 1973, "Magnitude-frequency analysis and channel morphology," Morisawa, M. (ed.), *Fluvial Geomorphology*, Allen and Unwin, London, UK.

Ely, L.L., et al., 1993, "A 5000-year record of extreme floods and climate change in the southwestern United States," *Science* 262:410–412.

Ely, L.L., Webb, R.H., Enzel, Y., 1992, "Accuracy of post-bomb ^{137}Cs and ^{14}C in dating fluvial deposits," *Quaternary Research* 38:196–204.

England, J.F., Jr., Salas, J.D., Jarrett, R.D., 2003a, "Comparisons of two moments-based estimators that use historical and paleoflood data for the log-Pearson Type III distribution," Water Resources Research 39(9):SWC-5-1 through SWC-5-16, doi:10.1029/2002WR00179.

England, J.F., Jr., Jarrett, R.D., Salas, J.D., 2003b, "Data-based comparisons of moments estimators that use historical and paleoflood data," *Journal of Hydrology* 278:170–194.

England, J.F. Jr., et al., 2010, "Paleohydrologic bounds and extreme flood frequency of the Arkansas River Basin, Colorado, USA," *Geomorphology* 124:1–16, doi:10.1016/j.geomorph.2010.07.021.

Enzel,Y., et al., 1993, "Paleoflood evidence for a natural upper bound to flood magnitudes in the Colorado river basin," *Water Resources Research* 29, 2287–2297.

Fanok, S.F., and Wohl, E.E., 1997, "Assessing the accuracy of paleohydrologic indicators, Harpers Ferry, West Virginia," *Journal of the American Water Resources Association* 33:1091–1102.

Harden, T.M., et al., 2011, "Flood-frequency analyses from paleoflood investigations for Spring, Rapid, Boxelder, and Elk Creeks, Black Hills, western South Dakota," Scientific Investigations Report 2011-5131, U.S. Geological Survey, Reston, VA.

Hosman; K.L., Ely, L.L., O'Connor, J.E., 2003, "Holocene paleoflood hydrology of the Lower Deschutes River, Oregon," O'Connor, J.E., Grant, G.E. (eds.), *A Peculiar River: Geology, Geomorphology, and Hydrology of the Deschutes River, Oregon*, Water Science and Application 7, American Geophysical Union, Washington, DC.

House, P.K. and Baker, V.R., 2001, "Paleohydrology of flash floods in small desert watersheds in western Arizona," *Water Resources Research* 37:1825–1839.

House P.K., et al. (eds.), 2002, *Ancient floods, modern hazards: Principles and Applications of Paleoflood Hydrology*, Water Science and Application Series, Vol. 5, American Geophysical Union, Washington, DC.

Jacoby, Y., et al., 2008, "Late Holocene upper bounds of flood magnitudes and twentieth century large floods in the ungauged, hyperarid alluvial Nahal Arava, Israel," *Geomorphology* 95:274–294.

Jarrett, R.D., and England, J.F., Jr., 2002, "Reliability of paleostage indicators for paleoflood studies," House, P.K., et al. (eds.), *Ancient Floods, Modern Hazards: Principles and Applications of Paleoflood Hydrology*, Water Science and Application Series, Vol. 5, American Geophysical Union, Washington, DC.

Kidson, R.L., Richards, K., Carling, P.A., 2005, "Hydraulic model calibration for extreme floods in bedrock-confined channels: case study from northern Thailand, *Hydrologic Processes* 20:329–344.

Knox, J.C., 1999, "Long-term episodic changes in magnitudes and frequencies of floods in the Upper Mississippi Valley," Brown, A.G., and. Quine, T.A (eds.), *Fluvial Processes and Environmental Change*, John Wiley & Sons, New York, NY.

Knox, J.C., 2000, "Sensitivity of modern and Holocene floods to climate change," *Quaternary Science Reviews* 19:439–457.

Knox, J.C., and Daniels, J.M., 2002, "Watershed scale and the stratigraphic record of large floods," House, P.K., et al. (eds.), *Ancient Floods, Modern Hazards: Principles and Applications of Paleoflood Hydrology*, Water Science and Application Series, Vol. 5, American Geophysical Union, Washington, DC.

Kochel, R.C., 1980, "Interpretation of flood paleohydrology using slackwater deposits, Lower Pecos and Devils Rivers, Southwestern Texas," Ph.D. dissertation, University of Texas, Austin, TX.

Kochel, R.C., and Baker, V.R., 1982, "Paleoflood hydrology," *Science* 215:353–361.

Kochel, R.C., and Baker, V.R., 1988, "Paleoflood analysis using slack water deposits," Baker, R.V., Kochel, R.C., Patton P.C. (eds.), *Flood Geomorphology*, John Wiley & Sons, New York, NY.

Levish, D.R., England, J.F. Jr., Klawon, J.E. and O'Connell, D.R.H., 2003, "Flood Hazard Analysis for Seminoe and Glendo Dams, Kendrick and North Platte Projects, Wyoming," Final Report, Bureau of Reclamation, Denver, CO, November, 126 p. and two appendices.

O'Connell, D.R.H., 2005, "Nonparametric Bayesian flood frequency estimation," *Journal of Hydrology* 313:79–96.

O'Connell, D.R.H., et al., 2002, "Bayesian flood frequency analysis with paleohydrologic bound data," *Water Resources Research* 38:1058. doi:10.1029/2000WR000028.

O'Connor, J.E., Webb, R.H., 1988, "Hydraulic Modeling for Palaeoflood Analysis," Baker, R.V., Kochel, R.C., Patton P.C. (eds.), *Flood Geomorphology*, John Wiley & Sons, New York, NY.

O'Connor, J.E., et al., 1994, "A 4500-year record of large floods on the Colorado river in the Grand Canyon, Arizona," *The Journal of Geology* 102:1–9.

Reis, D.S., Jr., Stedinger J.R., 2005, "Bayesian MCMC flood frequency analysis with historical information," *Journal of Hydrology* 313:97–116.

Ruiz-Villanueva, V., et al., 2010, "Dendrogeomorphic analysis of flash floods in a small ungauged mountain catchment (Central Spain)," *Geomorphology* 118:383–392.

Schumm, S. A., 1968, "River adjustment to altered hydrologic regimen - Murrimbidgee River and paleochannels," U.S. Geological Survey Professional Paper 598, Reston, VA.

Sigafoos, R.S., 1964, "Botanical evidence of floods and flood-plain deposition," U.S. Geological Survey Professional Paper 485A, Reston, VA.

Springer, G.S., and Kite, J.S., 1997, "River-derived slackwsater sediments in caves along Cheat River, West Virginia," *Geomorphology* 18:91–100.

Stedinger, J.R., and Cohn, T.A., 1986, "Flood frequency analysis with historical and paleoflood information," *Water Resources Research* 22:785–793.

Thorndycraft, V., et al., 2005a, "A long-term flood discharge record derived from slackwater flood deposits of the Llobregat River, NE Spain," *Journal of Hydrology* 313:16–31.

Webb, R.H., and Jarrett, R.D., 2002, "One-dimensional estimation techniques for discharges of paleofloods and historical floods," House, P.K., et al. (eds.), *Ancient Floods, Modern Hazards: Principles and Applications of Paleoflood Hydrology*, Water Science and Application Series, Vol. 5, American Geophysical Union, Washington, DC.

Wohl, E.E., 2002, "Modeled paleoflood hydraulics as a tool for interpreting bedrock channel morphology," House, P.K., et al. (eds.), *Ancient Floods, Modern Hazards: Principles and Applications of Paleoflood Hydrology*, Water Science and Application Series, Vol. 5, American Geophysical Union, Washington, DC.

Yanosky, T.M., and Jarrett, R.D., 2002, "Dendrochronologic evidence for the frequency and magnitude of paleofloods," House, P.K., et al. (eds.), *Ancient Floods, Modern Hazards: Principles and Applications of Paleoflood Hydrology*, Water Science and Application Series, Vol. 5, American Geophysical Union, Washington, DC.

3.6.3 Hydrologic Hazard Methods for Dam Safety

John F. England, Jr.[1]

[1]U.S. Department of Interior, Bureau of Reclamation, Flood Hydrology and Consequences Group, 86-68250, Technical Service Center, Denver, CO 80225

Historically, dam design and analysis methods have focused on selecting a level of protection based on spillway evaluation flood loadings. Traditionally, the protection level is based on the probable maximum flood (PMF). Reclamation uses risk analysis to assess the safety of dams, recommend safety improvements, and prioritize expenditures (Reclamation, 2011). Risk analysis, from a hydrologic perspective, requires an evaluation of a full range of hydrologic loading conditions and possible dam failure mechanisms tied to consequences of a failure. This risk approach is in contrast to the traditional approach of using single upper bound, maximum events such as the PMF which, from a risk perspective, have no probability of occurrence and resulting potential consequences are not clearly defined.

The flood loading input to a dam safety risk analysis within Reclamation is a hydrologic hazard curve that is developed from a hydrologic hazard analysis (HHA). Hydrologic hazard curves (HHC) are peak flow and volume probability relationships. These hazard curves are presented as graphs and tables of peak flow and volume (for specified durations) versus annual exceedance probability (AEP). The range of AEPs displayed on these graphs is intended to be sufficient to support the decision making needs of the organization.

The hydrologic load inputs to a risk analysis may consist of peak flows, hydrographs and reservoir levels and their annual exceedance probabilities. From a prioritization of resources perspective, it is difficult to accomplish dam safety goals from a "true or false" assessment of whether a structure is capable of withstanding a probable maximum flood. Therefore, a single, deterministic flood estimate such as the PMF is no longer adequate to evaluate the hydrologic safety of a dam. As an alternative, Reclamation has approached prioritization and modification needs through estimates of flood magnitudes and volumes and their associated exceedance probabilities up to the PMF. This is a risk analysis approach.

Hydrologic hazard curves provide magnitudes and probabilities for the entire ranges of peak flow, flood volume (hydrograph), and reservoir elevations, and do not focus on a single event. The peak flow, flood volume, and maximum reservoir level frequency distributions for dams with potentially high loss of life might extend to very low probabilities. For dam safety risk assessments, flood estimates are needed for AEPs less than 1 in 10,000 (1×10^{-4}) to satisfy Reclamation's public protection guidelines (Reclamation, 2011).

Reclamation has developed and applied numerous methods to estimate hydrologic hazard curves—extreme flood magnitudes and probabilities for dam safety. These methods can be broadly classified into streamflow-based statistical approaches, and rainfall-based (with runoff) statistical approaches. Current hydrologic hazard curve methods used by Reclamation are summarized in Table 3–1, and are generally ranked according to the level of effort involved. Methods are described in Swain et al. (2006); improvements to these current methods and other tools and approaches are ongoing and may be added as project needs and experience dictates. Recent comparisons in California and Colorado have shown relatively close agreement betweenstochastic rainfall-runoff models and paleoflood-based frequency curves in some detailed study cases. The methods used and some principles involved in estimating hydrologic hazards are presented.

Table 3–1. Current Methods Used by Reclamation To Develop Hydrologic Hazard Curves

Class	Method of Analysis and Modeling (*reference*)	Hydrologic Hazard Curve Product	Risk Analysis/ Design Level[1]	Level of Effort[2]
Streamflow-based statistics	Peak-flow frequency analysis with historical/paleoflood data - Graphical method (*Swain et al., 2006*)	peak flow frequency, volume frequency; hydrographs	CFR, IE	Low
Streamflow-based statistics	Peak-flow frequency analysis with historical/paleoflood data - EMA (*Cohn et al., 1997*)	peak flow frequency	IE, CAS, FD	Low
Streamflow-based statistics	Peak-flow frequency analysis with historical/paleoflood data - FLDFRQ3 (*O'Connell et al., 2002*)	peak flow frequency	IE, CAS, FD	Low
Streamflow-based statistics	Hydrograph Scaling and Volumes (*England, 2003*)	hydrographs and volumes; based on peak flow frequency	CFR, IE, CAS, FD	Low
Rainfall-based statistics and Runoff Transfer	GRADEX Method (*Naghettini et al., 1996*)	volume frequency; hydrographs	IE, CAS, FD	Moderate
Rainfall-based statistics and Rainfall-Runoff	Australian Rainfall-Runoff Method (*Nathan and Weinmann, 2001*)	peak flow and hydrographs; based on rainfall frequency and PMP	IE, CAS, FD	Moderate
Rainfall-based statistics and Rainfall-Runoff	Stochastic Event-Based Precipitation Runoff Modeling with SEFM (*Schaefer and Barker, 2002*)	peak flow frequency; hydrographs; volume frequency; reservoir elevation frequency	CAS, FD	High
Rainfall-based statistics and Rainfall-Runoff	Stochastic Rainfall-Runoff Modeling with TREX (*England et al., 2006, 2007*)	peak flow frequency; hydrographs; reservoir elevation frequency	CAS, FD	High

1   CFR: Comprehensive Facility Review; IE: Issue Evaluation; CAS: Corrective Action Study; FD: Final Design

2   Low: less than 20 staff days; Moderate: 21 to 100 staff days; High: more than 100 staff days

References

Bauer, T.R. and Klinger, R.E., 2010. Evaluation of Paleoflood Peak Discharge Estimates in Hydrologic Hazard Studies. Report DSO-11-03, Dam Safety Technology Development Program, Bureau of Reclamation, Denver, 19 p.

England, J.F., Jr. (2003), "Probabilistic Extreme Flood Hydrographs that Use Paleoflood Data for Dam Safety Applications," Bureau of Reclamation, Denver, CO, January 2003, http://www.usbr.gov/ssle/damsafety/TechDev/DSOTechDev/DSO-03-03.pdf

England, J.F., Jr. (2004), "Review of Selected Large Flood Estimates in the United States for the U.S. Geological Survey," Bureau of Reclamation, Denver, CO, August, 14 p.

England, J.F., Jr. (2011a), "Hydrologic Hazard Analysis," Section 3, *Dam Safety Risk Analysis Best Practices Training Manual*, Bureau of Reclamation, Denver, CO, http://www.usbr.gov/ssle/damsafety/Risk/BestPractices/03-HydrologicHazard201108.pdf

England, J.F., Jr. (2011b), "Flood Frequency and Design Flood Estimation Procedures in the United States: Progress and Challenges," *Australian Journal of Water Resources* 15(1):33-46, Institution of Engineers, Australia.

England, J.F., Jr., and Swain, R.E. (2008), "Extreme Flood Probability Estimation for Dam Safety," 28th USSD Annual Conference, April 28–30, 2008, United States Society on Dams, Portland, OR, 12 p.

England, J.F. Jr., Klawon, J.E., Klinger, R.E. and Bauer, T.R. (2006) Flood Hazard Study, Pueblo Dam, Colorado, Final Report, Bureau of Reclamation, Denver, CO, June, 160 p. and seven appendices.

Levish, D.R., England, J.F. Jr., Klawon, J.E. and O'Connell, D.R.H. (2003) Flood Hazard Analysis for Seminoe and Glendo Dams, Kendrick and North Platte Projects, Wyoming, Final Report, Bureau of Reclamation, Denver, CO, November, 126 p. and two appendices.

Reclamation (1999) A Framework for Characterizing Extreme Floods for Dam Safety Risk Assessment. Prepared by Utah State University and Bureau of Reclamation, Denver, CO, November, 67 p.

Reclamation (2002a) Interim Guidelines for Addressing the Risk of Extreme Hydrologic Events. Bureau of Reclamation, Denver, CO, August, 3 p.

Reclamation (2002b) Flood Hazard Analysis - Folsom Dam, Central Valley Project, California. Bureau of Reclamation, Denver, CO, January, 128 p. and 4 appendices

Reclamation (2011), "Dam Safety Public Protection Guidelines, A Risk Framework to Support Dam Safety Decision-Making," Bureau of Reclamation, Denver,CO, http://www.usbr.gov/ssle/damsafety/documents/PPG201108.pdf.

Swain, R.E., England, J.F. Jr., Bullard, K.L. and Raff, D.A. (2006) Guidelines for Evaluating Hydrologic Hazards, Bureau of Reclamation, Denver, CO, 83 p.

Vogel, R.M., Matalas, N.C., England, J.F. and Castellarin, A. (2007), "An assessment of exceedance probabilities of envelope curves," Water Resources Research 43 (7), W07403, doi:10.129/2006WR005586.

PANEL 3

EXTREME PRECIPITATION EVENTS

Co-Chairs:
John England, BoR
Chandra Pathak, USACE

Technical Reporters:
Nebiyu Tiruneh and Brad Harvey, NRC

4. Extreme Precipitation Events

John England[1], Chandra Pathak[2], Nebiyu Tiruneh[3] and Brad Harvey[3]

[1]Bureau of Reclamation, Denver, CO
[2]U.S. Army Corps of Engineers, Washington, DC
[3]Office of New Reactors, U.S. NRC, Rockville, MD

4.1 Motivation

While there are many drivers of extreme floods, Session 3 of the Probabilistic Flood Hazard Assessment (PFHA) Workshop focused on extreme precipitation events, such as those from tropical cyclones, atmospheric rivers, and locally intense convective rainfall. Extreme storm precipitation is also a major input to rainfall–runoff models that are used to estimate extreme floods. Traditional data and methods for estimating extreme storms for nuclear facilities and dam safety that typically focus on the probable maximum precipitation (PMP) are outdated. Currently, there is no mechanism in place within Federal agencies to routinely collect, analyze, and archive extreme storm data or to research and develop improved techniques useful for estimating extreme floods and their probabilities.

Over the past 25 years, there have been major advances in precipitation data collection and processing (such as radar data), extreme precipitation analysis and statistics, and physically based modeling of extreme storms. There has been much work on regional precipitation frequency analysis of extremes and applications for dam safety. These efforts and advances in meteorology and hydrometeorology of extremes are crucial and are being used by some agencies as inputs to probabilistic flood hazard assessments for critical infrastructure.

4.2 Background

Traditionally, extreme precipitation estimates for nuclear facilities and dam safety have focused on the PMP. However, the PMP data sets and procedures have not kept pace with changes in data, technology, and the need for more comprehensive information beyond maximum estimates, including full probability distributions of extreme precipitation. In addition, there are no widely accepted procedures in place to update storm databases, methodology, and reports that are used to estimate extreme precipitation events useful for probabilistic flood hazard assessments.

Panel 3, "Extreme Precipitation Events," highlighted recent data sets, improved methods, and applications for estimating extreme precipitation events, probabilities, and numerical modeling of extremes. The focus was to present and discuss these numerous advances in observations, data, statistics, and modeling, and to highlight ongoing applications, research, and collaborative opportunities for making advances in extreme precipitation estimates and probabilities.

4.3 Overview of Presentations

Daniel Wright (Princeton University) presented observations and methods for rainfall and flood frequency using high-resolution radar rainfall fields and stochastic storm transposition. The work focused on urban areas; he highlighted the differences between hydrologic processes that govern the hydrology in urban vs. non-urban areas. The work is based on high-resolution radar and stochastic storm transposition (SST). A recap of the conventional hydrologic methods was presented. Work includes area reduction factors (ARF) for tropical vs. non-tropical storms. Daniel used a case study on a watershed in Charlotte, NC to illustrate the storm rainfall and probabilities with the Gridded Surface Subsurface Hydrologic (GSSHA) Model. He described future research areas for SST applications including reservoir operating rules, combining with PMP/PMF, improving flood risk estimation in data-poor regions, and to account for climate change.

Mel Schaefer (MGS Engineering) gave a detailed presentation on regional precipitation frequency analysis and extremes including PMP, focusing on practical considerations. He highlighted several key areas, including: PMP being just another point on the frequency curve; the benefits of regional analyses, trading space for time sampling, and dramatically reducing uncertainties; spatial mapping of precipitation with isopercental method; and Monte Carlo simulation. The methods include large data sets, use of L-moments, PRISM datasets, spatial mapping, and Monte Carlo simulation. Annual exceedance probabilities (AEPs) for near-PMP events have a much wider range than assumed in the engineering community.

R. Jason Caldwell (Reclamation) presented research completed by Kelly Mahoney (CIRES/NOAA) on using high-resolution numerical modeling as a tool to assess extreme precipitation events. The key research question was: Can we use knowledge of past events and high-resolution numerical modeling to improve PMP estimates? She used the Weather Research and Forecasting (WRF) model and illustrated a case study in western Colorado. The approach was to use current data on select watersheds and perturb using numerical models to maximize an individual storm. Control simulations were run, and then perturbed. Control simulation involved moisture maximization by increasing relative humidity (RH) by a specified percentage. A climate-change perturbation was performed using two climate models.

Geoff Bonnin (NWS) gave an overview and perspective on precipitation frequency estimates for the nation and extremes. A status of NOAA Atlas 14 (Precipitation Frequency Atlas of the United States) was given to illustrate the volumes, Average Recurrence Intervals, and durations (NOAA/NWS, 2004). Geoff explained how terms, semantics, etc., related to extremes and climate change requires better understanding (Bonnin et al, 2011). Better guidance on potential impacts of climate change on intensity frequency duration (IFD) curves is needed (NAS, 2011). He also raised and discussed the question, "Can climate change make the 'perfect storm' more perfect?"

Victoria Sankovich (Reclamation) gave a perspective on extreme precipitation frequency for dam safety and nuclear facilities. She presented four case studies, illustrating varying complexity of data and methods, for locations in North Carolina, New Mexico, Idaho and Oklahoma (Kallio and Klinger, 2012; Wright and Sankovich, 2011; England et al., 2012; Novembre et al., 2012). Data sets varied and included NOAA 14, point precipitation data, and newer gridded data sets from the NOAA Climate Forecast System Reanalysis (CFSR). Methods included PMP probability estimates from Australian Rainfall-Runoff (ARR) and regional precipitation frequency with L-Moments. Evolving ideas include radar-derived rainfall for spatial patterns and temporal patterns (mass curves) from modeled data in CFSR. The extreme precipitation frequency data and methods are a scaled approach based on dam and consequences.

4.4   Summary of Panel Discussion

The panel members discussed questions proposed by the moderators and audience:

1. Describe the advancements and improvements in extreme storm rainfall and precipitation observations and databases over the past 30 years. Are there opportunities with radar, point observations, reanalysis data sets, and other data that can readily be used for extreme precipitation analyses, understanding, and applications for critical infrastructure?

Thirty additional years of rainfall data is useful. Radar data are now also useful in providing a better picture of spatial and temporal correlation of rainfall data for numerical weather models. Caution should be used in interpreting radar data; rainfall for intense events can be underestimated and the useful range of radar data can be overestimated. Based on records of the past 20 to 25 years, we are learning about the capabilities and limitations of radar, including orographic effects.

2. Outline the advances in statistical and data-processing methods that can be used for extreme precipitation frequency estimation. These might include regional precipitation frequency, regionalization of parameters, Geographic Information Systems, climatological estimation (such as PRISM), and other areas. How might these tools be applied in practice, and how might they include uncertainty estimates?

Fitting generalized extreme value (GEV) distribution to bound shape parameters might help with improvements in PMF. Changes in regional analysis are also important; for example, mapping a large heterogeneous region into smaller multiple homogeneous regions (paying attention to developing criteria that establish regional boundaries) is a technique for handling spatial diversity. Another technique would be to account for seasonal variability in evaluating the climatology and meteorological physics of rainfall events. The shortage of funds has limited the widespread application of research; although the results obtained so far by the research community working in small groups here in the United States is encouraging.

Related Question: Confidence intervals (CIs) (for example, as shown in Geoff Bonnin's presentation) are great in indicating the statistical level of confidence; however, CIs do not provide practical solutions and meaningful interpretations beyond that. Should research be focusing on how CIs are to be used to address uncertainties? There is a potential for improving flood forecasting models using recent advances and better understanding of statistical parameters and measures such as CIs.

3. Describe the advances in physical and numerical modeling of extreme precipitation (such as the Weather Research and Forecasting Model, WRF) that can give insights into the processes and magnitudes of extreme precipitation, including spatial and temporal distributions. How can these tools be applied to provide practical limits to extreme precipitation magnitudes, spatial and temporal storm patterns, transposition, and other extreme storm scaling?

Improving the analysis of extreme precipitation events is difficult simply because such events are so rare and they require high-resolution observing systems to be in place in order to gather appropriate observations. Higher-resolution models can potentially provide better results and avoid the need for manual processes such as storm transposition. It is important to identify two issues in this respect: (1) from a science perspective, models are useful for testing hypotheses; and (2) it is necessary to improve model performance using data, new methods, and experience. Numerical weather models can be useful in improving our understanding of the physics and the characteristics and limits of extreme precipitation events, but more work needs to be done to improve the performance of the models and to demonstrate that the models do represent reality. Shortage of funding continues to restrict research, improvement to models, and advances in techniques. Caution needs to be exercised on parameter selection and estimation. Choice of parameters should not be done for the sole purpose of reproducing records; it is important to understand the physics. Because different researchers have different interpretations and approaches, forming ensembles requires care and a meticulous approach. More detailed information can be obtained from radar data and the use of numerical models such as WRF. It is important to remember the recent May 2010 Nashville flooding, which was the result of a stalled tropical air system. Therefore, careful thinking is required for storm transposition and weather research and forecasting (WRF) models have important practical application.

4. The National Research Council report on extreme precipitation (NAS, 1994) suggested research in several areas, including: radar hydrometeorology and storm cataloging, numerical modeling of extreme storms in mountainous regions, and estimating probabilities of extreme storm rainfalls. Are there existing technical barriers to fully probabilistic extreme storm estimation for assessing critical infrastructure, as opposed to Probable Maximum Precipitation?

Technical complexities include (1) size of watershed, (2) combining different types of storms (e.g., tropical vs. non-tropical), and (3) combining storms with snow melt. Better modeling results are computationally intensive. Based on experience in radar data application, it is a combination of access to computing resources and availability of the necessary skilled technical persons. Advances have also been restricted by lack of funds. The 2011 Mississippi river flood, which was the result of three months of long-duration rain storms and snowmelt, exhibited a 1 in 250 annual exceedance probability. This and other events like it affirm the need for improvements and updates of existing methods and techniques. Funding has been a major issue restricting such efforts.

Additional public questions and/or comments were as follows.

A gentleman with experience in radar data application and weather forecasting made the following comment: Numerical weather models are lagging behind; for example, WRF was not intended to analyze extreme rainfall events. He disagreed with Geoff Bonnin on the integration and limitations of radar data; NEXRAD could be used successfully for extreme events with the proper quality control and supplemental rain gauge data.

Panel responses were as follows. WRF is an appropriate tool for analyzing extreme events. WRF application for a couple of case studies in Colorado was encouraging, but questions still remain on how it could be best used for developing area-intensity relations.

How do you define or identify data independence when rainfall data are collected at numerous locations in a homogeneous watershed? The issue of using these data to define equivalent record lengths requires further examination.

Oversampling can be avoided by only using a subset of the collected data. Sorting extreme readings above a certain threshold by event date also helps identify situations in which the same storm is causing extreme readings at multiple locations. Understanding your watershed (e.g., rural vs. urban) is also important. The approaches used in Atlas 14 for Alaska and California could shed some light in this matter.

4.5   Observations and Insights

1. Opportunities in Extreme Rainfall Observations/Databases
   - point rainfall data and areal reduction factors
   - radar data – significant use; better spatial and temporal correlations
   - need extreme storm catalogs

2. Advances in statistics and data processing methods
   - regionalization techniques
   - storm spatial and temporal patterns
   - mapping larger regions, accounting for seasonal variability
   - uncertainty estimates

3. Physical and Numerical Modeling
   - radar and better resolution models provide better results
   - use models for hypothesis testing
   - evaluate past events (September 1970; May 2010 Nashville)
   - lack of funding restricts research

4. Technical and Other Barriers
   - technical complexities (watershed size, different storm mechanisms…)
   - computing resources
   - skilled personnel
   - funding

4.6   References

Bonnin, G. M., K. Maitaria, and M. Yekta, 2011, "Trends in Rainfall Exceedances in the Observed Record in Selected Areas of the United States," *JAWRA Journal of the American Water Resources Association* 47(6):1173–1182.

England, J.F., Jr., F. J. Dworak, and V.L. Sankovich, 2012, "Hydrologic Hazard Analysis, Anderson Ranch Dam, Boise Project, Idaho, Final Report," U.S. Department of the Interior, Bureau of Reclamation, Denver, CO.

Kallio, R., and R.E. Klinger, 2012, "Star Fort Dam Hydrologic Hazard, Ninety Six National Historic Site, South Carolina, National Parks Service," U.S. Department of the Interior, Bureau of Reclamation, Denver, CO.

Mahoney, K., et al., "High-resolution downscaled simulations of warm-season extreme precipitation events in the Colorado Front Range under past and future climates," *Journal of Climate*, in review.

National Academy of Sciences (NAS), 1994, *Estimating Bounds on Extreme Precipitation Events: A Brief Assessment*, National Research Council, National Academies Press, Washington, DC.

NAS, 2011, *Global Change and Extreme Hydrology: Testing Conventional Wisdom*, National Research Council, National Academies Press, Washington, DC, http://dels.nas.edu/resources/static-assets/materials-based-on-reports/reports-in-brief/Extreme-Hydro-Brief-Final.pdf

National Oceanic and Atmospheric Administration/National Weather Service (NOAA/NWS), 2004, *NOAA Atlas 14: Precipitation-Frequency Atlas of the United States*, Silver Spring, MD, http://www.nws.noaa.gov/oh/hdsc/currentpf.htm.

Novembre, N.J., et al., 2012, "Altus Dam Hydrologic Hazard and Reservoir Routing for Corrective Action Study," U.S. Department of the Interior, Bureau of Reclamation, Denver, CO.

Wright, J., and V.L. Sankovich, 2011, "Trapped Rock Dam Hydrologic Hazard, Zuni Pueblo Indian Reservation, NM, Bureau of Indian Affairs," U.S. Department of the Interior, Bureau of Reclamation, Denver, CO.

4.7   Abstracts

The workshop organizing committee developed the flooding topics to be discussed and chose the panel co-chairs for each panel topic. The co-chairs then identified potential speakers and discussed them with the organizing committee. Following agreement, the co-chairs sent out invitations to the presenters requesting presentation titles and abstracts for documentation in the workshop program. The speakers and panelists are identified in the workshop agenda (please see Appendix A). The following four abstracts document these presentations and, in some cases, reflect the discussions during the panel session.

4.7.1   High-Resolution Numerical Modeling as a Tool to Assess Extreme Precipitation Events

Kelly Mahoney

University of Colorado/Cooperative Institute for Research in the Environmental Sciences (CIRES) and NOAA Earth Systems Research Lab (ESRL), Boulder, CO

Weather and climate data is often downscaled to supply estimates, predictions, or projections to stakeholders and decisionmakers at higher, more useful resolution than that provided by contemporary observations and reanalysis datasets. The details of downscaling (e.g., model resolution, domain selection, and model physics) become of critical importance when addressing extreme weather phenomena at a local scale, because such events are often determined by (and/or sensitive to) small-scale processes. Particularly for flood events in complex terrain, it is necessary to resolve the details of both the terrain itself and the driving fine-scale atmospheric processes to form a realistic picture of future flood risk. This discussion will briefly summarize work that uses high-resolution (1-km) simulations of warm-season intense precipitation events in key geographic regions in both the western United States and the southeastern United States. The overall objective of the work is to improve understanding of the factors and physical processes responsible for both historical (observed) extreme events and future possible extreme events. In the western United States, historical events such as the Fort Collins, CO, flood of July 1997 are examined as well as select historical events deemed critical to a particular dam's planning purposes. In the Southeast United States, work is underway to explore the use of high-resolution modeling of key cases (e.g., the Tennessee floods of May 2010 and the Atlanta-area floods of September 2009) for future use in both forecast improvement and better hydrologic-response predictions. High-resolution numerical modeling can be a useful tool to address questions related to storm maximization methods, probable maximum precipitation (PMP) limits, and elevation adjustment strategies—concepts commonly used for water resources management nationwide. For example, is a storm of PMP magnitude physically able to develop in a specific region? In regions where paleoflood data do not suggest that floods of such magnitude have actually occurred historically, can a model be "forced" to produce such a storm following reasonable maximization of key fields? Do such changes reflect realistic possibilities of future climate change? Additional potential applications for decisionmaking in the realm of water resources management will be discussed.

4.7.2 An Observation-Driven Approach to Rainfall and Flood Frequency Analysis Using High-Resolution Radar Rainfall Fields and Stochastic Storm Transposition

Daniel Wright

Department of Civil and Environmental Engineering, Princeton University, Princeton, NJ 08544

Spatial and temporal variability in extreme rainfall, and its interactions with land cover and the drainage network, is an important driver of flood response. However, "design storms," which are commonly used for flood risk assessment, are assumed to be uniform in space and either uniform or highly idealized in time. The impacts of these and other commonly made assumptions are rarely considered, and their impacts on flood risk estimates are poorly understood. We develop an alternate framework for rainfall frequency analysis that couples stochastic storm transposition (SST) with "storm catalogs" developed from a ten-year high-resolution (15-minute, 1-km^2) radar rainfall dataset for the region surrounding Charlotte, North Carolina, USA. The SST procedure involves spatial and temporal resampling from these storm catalogs to reconstruct the regional climatology of extreme rainfall. SST-based intensity-duration-frequency (IDF) estimates are driven by the spatial and temporal rainfall variability from weather radar observations, are tailored specifically to the chosen watershed, and do not require simplifying assumptions of storm structure. We are able to use the SST procedure to reproduce IDF estimates from conventional methods for small urban watersheds in Charlotte. We demonstrate that extreme rainfall can vary substantially in time and in space, with important flood risk implications that cannot be assessed using conventional techniques. SST coupled with high-resolution radar rainfall fields represents a useful alternative to conventional design storms for flood risk assessment, the full advantages of which can be realized when the concept is extended to flood frequency analysis using a distributed hydrologic model. A variety of challenges remain which complicate the application of SST to larger watersheds and more complex settings, but the technique nonetheless represents a robust, observation-based alternative for assessing flood risk.

4.7.3 Regional Precipitation-Frequency Analysis and Extremes Including PMP – Practical Considerations

M.G. Schaefer[1]

[1]MGS Engineering Consultants, Inc.

Precipitation-Frequency relationships are now being developed for watersheds for use in rainfall-runoff modeling of extreme floods extending to Annual Exceedance Probabilities (AEPs) of 10^{-5} and beyond. This capability is made possible by advancements in several technical areas including regional-frequency analysis, L-moment statistics, spatial analysis of storms such as isopercental analysis, and GIS-based spatial mapping using radar data and ground-based precipitation measurements. Methodologies have been developed to derive the precipitation-frequency relationship in a manner that accounts for uncertainties in the various contributing components and provides uncertainty bounds for the mean frequency curve. In the frequency context, Probable Maximum Precipitation (PMP) becomes just another value on the frequency curve, which can be exceeded given the large uncertainties in PMP estimation. Experience has been gained in application of regional frequency analysis and development of precipitation-frequency relationships at over 20 watersheds in semi-arid to humid climates throughout the western United States and British Columbia. This presentation will describe experience gained about the behavior of precipitation-frequency characteristics which will provide guidance and assist judgments in future applications for a range of climatic environments.

4.7.4 Extreme Precipitation Frequency for Dam Safety and Nuclear Facilities – A Perspective

Victoria L. Sankovich[1], R. Jason Caldwell[1] and John F. England, Jr.[1]

[1]Bureau of Reclamation, Technical Service Center, Flood Hydrology, Denver, CO 80225

Federally regulated dams and nuclear power plants are designed to withstand extreme storm rainfall events. Some Federal agencies design their structures to the Probable Maximum Precipitation (PMP; e.g., Prasad et al., 2011; USACE, 1984). Probable maximum precipitation is defined, theoretically, as the greatest depth of precipitation for a given duration that is physically possible over a given storm area at a particular geographical location at a certain time of year (HMR; WMO 2009, Hansen et al., 1982). Each of the hydrometeorological reports (joint efforts by the Bureau of Reclamation, National Weather Service, and Army Corps of Engineers, among others) have served as the basis for design precipitation estimates in flood design studies of critical infrastructure for several decades. These estimates are based on storm data that are outdated by up to 40 years. In addition, designing every dam to meet PMP may be a cost-prohibitive and unnecessary approach depending on the associated hazards downstream.

In recent years, the Bureau of Reclamation has been using an approach which incorporates regional precipitation frequency analysis (L-moments; Hosking and Wallis, 1997) and, at times, applies the PMP as an upper limit to the frequency curve (Nathan and Weinmann, 2001). Nonetheless, there are many valuable concepts from PMP (i.e., storm analysis, storm maximization, depth-area-duration analysis, storm transposition), which lend information to the decisionmaking process. For example, observed isohyetal patterns from storms are overlaid on the drainage basin of Federal dams to examine rainfall-runoff relationships. Temporal patterns of observed rainfall are input to hydrologic models to accurately represent the time distribution of precipitation. Recent research by the Bureau of Reclamation into extreme precipitation events using storm maximization concepts suggests that events from the past two decades have the potential to meet or exceed current PMP estimates (Caldwell et al., 2011; e.g., Hurricane Floyd in 1999 and Hurricane Fran in 1996).

Here, we outline the methodology applied at various scales of project requirements through the presentation of case studies and recent research efforts at Reclamation. Examples will include: application and extrapolation of existing precipitation frequency curves (e.g., England et al., 2012; and Wright et al., 2012a and b); calculation of regional precipitation frequency relationships and uncertainty using L-moments (e.g., England et al., 2012, and Novembre et al., 2012); development of spatial and temporal patterns as input to hydrologic models (Novembre et al., 2012); and the utility of gridded meteorological datasets in hydrologic hazard assessments (Caldwell et al., 2011).

References

Caldwell, R.J., Sankovich, V.L., and England, J.F., Jr. (2011), "Synthesis of Extreme Storm Rainfall and Probable Maximum Precipitation in the Southeastern U.S. Pilot Region" (for the U.S. Nuclear Regulatory Commission, Office of Nuclear Regulatory Research), Bureau of Reclamation, Denver, CO, December 2011.

England, J.F., Jr., et al. (2012), "Hydrologic Hazard Analysis of Anderson Ranch Dam, Idaho Study," Bureau of Reclamation, Denver, CO, May 2012.

Hansen, E.M., Schreiner, L.C., and Miller, J.F. (1982), "Application of Probable Maximum Precipitation Estimates, United States East of the 105th Meridian," Hydrometeorological Report No. 52, National Weather Service, National Oceanic and Atmospheric Administration, U.S. Department of Commerce, Silver Spring, MD, 168 p.

Hosking, J.R.M., and Wallis, J.R. (1997), *Regional Frequency Analysis - An Approach Based on L-Moments*, Cambridge University Press, Cambridge, UK.

Nathan, R.J., and Weinmann, P.E. (2001), *Estimation of Large to Extreme Floods: Book VI in Australian Rainfall and Runoff, A Guide to Flood Estimation*, The Institution of Engineers, Australia.

Novembre, N., V. Sankovich, R. Caldwell, J. Niehaus, J. Wright, R. Swain, and J. England, Jr. (2012), "Altus Dam Hydrologic Hazard and Reservoir Routing for Corrective Action Study," Bureau of Reclamation, Denver, CO, August 2012.

Prasad, R., Hibler, L.F., Coleman, A.F., and Ward, D.L. (2011) for the U.S. Nuclear Regulatory Commission, "Design-Basis Flood Estimation for Site Characterization at Nuclear Power Plants in the United States of America," NUREG/CR-7046, PNNL-20091, prepared by Pacific Northwest National Laboratory, Richland, WA, November 2011, Agencywide Documents Access and Management System (ADAMS) Accession No. ML11321A195.

U.S. Army Corps of Engineers (USACE, 1984), "HMR52 Probable Maximum Storm (Eastern United States), User's Manual," CPD-46, Davis, CA, April 1987.

World Meteorological Organization (WMO, 2009), "Manual on Estimation of Probable Maximum Precipitation (PMP)," WMO No. 1045, Geneva, Switzerland, 259 p.

Wright, J., and V.L. Sankovich, and R. Swain (2011a), "Trapped Rock Dam Hydrologic Hazard, Zuni Pueblo Indian Reservation, NM, Bureau of Indian Affairs," U.S. Department of the Interior, Bureau of Reclamation, Denver, CO.

Wright, J., V. Sankovich, and R. Swain (2011b), "Tufa Stone Dam Hydrologic Hazard, for the Department of Interior Bureau of Indian Affairs," U.S. Department of the Interior, Bureau of Reclamation, Denver, CO, September 2011.

PANEL 4

FLOOD-INDUCED DAM AND LEVEE FAILURES

Co-Chairs:
Tony Wahl, BoR
Sam Lin, FERC

Technical Reporters:
Jacob Philip, Hosung Ahn and Juan Uribe, NRC

5. Flood-Induced Dam and Levee Failures

Tony Wahl[1], S. Samuel Lin[2], Jacob Philip[3], Hosung Ahn[4], Juan Uribe[5], and Wendy Reed[3]

[1]Bureau of Reclamation, Hydraulics Laboratory, Denver, CO
[2]Federal Energy Regulatory Commission, Washington, DC
[3]Office of Nuclear Regulatory Research, U.S. NRC, Rockville, MD
[4]Office of New Reactors, U.S. NRC, Rockville, MD
[5]Office of Nuclear Reactor Regulation, U.S. NRC, Rockville, MD

5.1   Motivation

Failures of dams and levees can be a source of extreme flooding at downstream and protected areas. There are many potential triggers for dam and levee failures, including but not limited to: hydrologic events, seismic events, storm surge, tsunamis, operational incidents, etc. This panel considered only dam and levee failures initiated by potential failure modes tied to hydrologic events. The panel's objective was to focus on defining and discussing the current states of the art and practice along with research needs related to the probabilistic analyses, evaluation, and assessment of potential flood hazards caused by dam and levee failures. Intentionally, presenters emphasized concepts and frameworks to apply the current advantageous risk-informed decision making (RIDM) approach for the hydrologic safety of dams and levees.

5.2   Background

Traditionally, dams and levees have been built to withstand particular flood magnitudes with corresponding low probabilities of occurrence as determined by required design standards. If a larger flood were to occur, the dam or levee could be overtopped and washed out, causing the large amount of stored water behind it to be released. High water levels behind dams and levees can also increase the probability of embankment failure by internal erosion associated with seepage through an embankment and/or its foundation. This panel's purpose was to address methods for estimating probabilities of failure, making probabilistic assessments of flood hazards, and determining inflow design floods. This process requires information on extreme flood loading, structural response, and consequences of a dam or levee failure.

5.3   Overview of Presentations

David Bowles provided an overview of risk assessment methods for dam and levee failures, stressing the importance of thoroughly identifying hazards and failure modes and quantifying probabilities of failure and consequences (life-loss and economic). He introduced the concept of tolerable risk and the use of logic trees to enable the incorporation of epistemic uncertainty into the risk-assessment process. Dr. Bowles also provided examples of risk-reduction methods.

Timo Schweckendiek detailed the Dutch approach to analyzing dam and levee reliability. After significant flooding of the Netherlands in 1953, which resulted in over 1800 people losing their lives, risk-based analyses have been used to assess dam and levee safety. The Netherlands' Delta Works Committee set a reliability target (for Central Holland) of 1/125,000 or 8×10^{-6} annual probability of failure and has used risk-based approaches in designing storm surge protection barriers.

Jason Hedien described a risk-informed decisionmaking (RIDM) case study involving two FERC-regulated high-hazard dams. This marked the first foray of FERC into using the RIDM process. The purpose of the study was to determine the Inflow Design Flood (IDF) for those two dams. This was achieved through an initial potential failure mode analysis (PFMA) as required by FERC, followed by a risk assessment workshop. The result of the study was that the spillways for both dams should be enlarged to a certain degree to meet tolerable risk guidelines.

Tony Wahl told the audience how the BoR has traditionally calculated dam breach parameters using mathematical regression models and estimated probabilities of failure via engineering judgment (which could be inconsistent and is subjective). He discussed recent development of physically based dam breach models and indicated how these tools could be used in the future to estimate both breach parameters and probabilities of failure. He stated that it is now possible to measure erodibility of embankment dam materials and described how the erodibility is extremely dependent on the compaction of the dam materials (compaction energy and water content at time of compaction).

David Margo gave an overview of the USACE dam and levee safety program, describing how it has evolved into a risk-informed approach. The USACE has a decision-driven focus: The type and nature of the decision drives how the risk analysis is carried out. Life safety is paramount in the USACE's decisionmaking process. Mr. Margo described how the USACE has a holistic approach to risk assessment, in that it looks at how a dam performs as a system and asks, "What are the failure modes and the key elements that could contribute to failure?".

5.4   Summary of Panel Discussion

1.   The probabilistic flood hazard analysis (specifically) and probabilistic risk assessment (PRA) (generally) for dam and levee failures need a structured evaluation process like the Senior Seismic Hazard Analysis Committee (SSHAC)process. Because the specific analysis often relies on expert assessments which are biased by human errors, the PRA analysis should evaluate the comprehensive uncertainties of data and modeling.

2.   Flood frequency analysis in Bulletin 17B is empirical, being based primarily on observed flood data. In contrast, PFHA involves highly subjective processes in estimating probability of breach occurrences with limited historical records; therefore, the analysis should consider all potential subjectivities and uncertainties in formulating breach scenarios and the system's responses to them.

3.   There are not adequate data on large dam failures. Reliable data on dams, dam components, and operations are generally not available to meet specific needs of risk assessments for individual dams or even components of dam systems.

    a.   History shows that large dams seldom fail by flooding alone, but by a combination of factors as a system.

    b.   We need more reliable data on dam and dam-components failure, and operations.

    c.   Better data would enable us to reduce subjectivity significantly in risk analysis and decisionmaking.

4.   The estimation of the probability of dam and levee failure at different loading levels (fragility) is difficult and often subject to bias and limited by engineering tools.

5.   The use of risk analyses (conducting comprehensive PFHA) is impeded by the lack of engineers who are trained and are comfortable with the methods that address probability, systems engineering for dams and levees, and risk in high-hazard situations.

6.   Probability estimate and risk analysis for dams have come a long way in the past 25 years. Statistics on dam failure in 1975 (general conclusion as $P_{failure} = 1 \times 10^{-4}$) was not a welcome message to dam owners.

7.   It is generally believed that the chances of dam failure are low (still around 1×10^{-4}, not much change in 40 years), but failures continue to occur. Failures are rarely caused by one factor, but often the "uncommon combination of not uncommon events." This needs to be addressed through risk analysis of dam systems.

8.   In some cases (e.g., the Canadian and Taum Sauk dam failure events), event trees/fault trees cannot handle the complexity of events.

9.   There are a lot of problems with data regarding dam failure events and the performance of structures, systems and components:

   a.   In the past, engineers did not want to talk about failure.

   b.   Failure reports tend to only be produced under regulatory requirement.

   c.   We do not keep track of near-failure incidents which are useful to support risk analyses. (In case of failures, we need to look at how we can make the data useful and applicable in the future.)

   d.   However, some data are available:

      i.   RMC (Risk Management Center of USACE) may have some information on reliability of mechanical/electrical components of dams (USACE's Chris Schulz has a database on mechanical failure).

      ii.   In Canada, CEATI's (The Centre for Energy Advancement through Technological Innovation's) Dam Safety Interest Group may be another source of similar information.

      iii.   National Performance of Dams Program at Stanford has data on dam incidents.

10.   In regards to Tony Wahl's statement that there is an order-of-magnitude uncertainty in predicting time of failure, it was asked whether that meant, for instance, we have a range of 3 min. to 30 min. or 3 hr. to 30 hr., and why does it vary so much? The reply given was that it is an order of magnitude uncertainty around the predicted time of failure from a given regression equation for dam-break, so it depends on the size of the dam. A point was also made that one big source of uncertainty about failure times is lack of accuracy/consistency in eyewitness reports of dam failure times. Sensitivity to erodibility of soils and its sensitivity to compaction and construction processes is also a factor.

11.   A recent experience in lab dam breach tests has shown that time of failure (i.e., rate of erosion) for embankment dams is extremely sensitive to erodibility of soils, which is in turn sensitive to compaction processes and conditions during original construction. Tests are now available that can evaluate soil erodibility (in the laboratory or in the field).

12.   Historically, dams have done well. The experience worldwide looks good.

13.   Tools have recently become available and continue to be developed that will enable us in the future to add probability and uncertainty to more nodes in the event trees describing sequences of events that lead to dam failure (e.g., process-based erosion and breach models).

14.   Data are not perfect, but sufficiently good to have confidence in the analyses performed. However, statistical analyses should move toward Bayesian methods to cope with the quality of historical data and the uncertainty in data analysis. This method has its advantages, especially in defining the confidence of the estimates statistically.

15.   The tolerable risk that society accepts for dams is less than the risk society accepts in other arenas. Improved guidance for tolerable risk limits for dams is needed.

16.   The state of PRA in dam and levee safety is relatively new compared to other fields, such as nuclear power plants, where a 2nd or 3rd generation of PRA is already underway.

17.   Dam failure analysis has been focused heavily on geotechnical engineering practices, but should also incorporate structural, hydrologic, and hydraulic engineering aspects.

18.   A major problem occurs when applying prescriptive methodologies in a "design-safe" manner. This prohibits thinking "outside the box" and hinders advancement in the field. On the other hand, it is costly, risky and challenging to analyze and tackle challenges in a manner outside the normal convention.

5.5   Observations and Insights

The following observations and insights were made in addition to the above addressed overview of presentations and panel discussions.

1.   The risk-informed approach is used throughout the federal community in a need-specific, inconsistent manner. Therefore, in order to improve risk assessment within Federal agencies, some areas for improvement are:

        a.   Improve failure-mode identification criteria

        b.   Try to achieve better scoping of risk assessments and risk model formulation

        c.   Understand the time scale (e.g., initiating, peaking, ending, duration, etc.) of flood causing and dam failure events and the time effects of specific events within a risk assessment.

        d.   Try to better consider uncertainties.

2. In the Netherlands, the risk analysis field gained popularity after a major flood in 1953; therefore it was event-driven and historically grown. A series of levees were installed to reduce the likelihood of flooding disasters.

3. Risk analysis in the Netherlands is performed by analyzing units, where each unit is composed of dike rings and polders (the term used to describe any piece of land reclaimed from water). These units are linked together and the longer the link, the more likely they are to fail. When analyzing these units, they can be broken down into smaller segments to produce a more homogenous section, and to simplify modeling analysis.

4. In the Netherlands, the reliability is determined from physically based computational models and expert judgment. The key is to account for correlations and dependencies. Failure probability results in the order of 10^{-6}.

5. If a dam cannot pass the PMF, then it should be evaluated against the inflow design flood and successfully pass. The dam failure analysis must also take into account the loss of life—NRC staff commented later that the dam failure analysis must be done independently (regardless of passing the PMF or not) and jointly (in combination) with the PMF and other flooding events as specified in the Section 2.4.4 of the NRC's Standard Review Plan, and that the NRC dam failure accounts for not the loss of life but the risk of flooding to plants and their safe operation.

6. When risk must be reduced at a certain dam, consideration should be given to infrastructure enhancements, such as upgrading an existing spillway, adding a new spillway, and/or armoring the embankment, among other things.

7. It is also a key to verify that risk-reduction measures performed on upstream dams (e.g., modifications, alterations, etc.) do not adversely affect downstream conditions.

8. Probability and risk analyses have been impeded by engineering educations in which topics on probability and risk are not emphasized.

9. Currently, the chances of dam failures are extremely low due to higher safety standards and the development of new design and construction technologies, but the failures still occur by the "uncommon combinations of not uncommon factors," often with identical failure sequences. This should be handled by risk analysis.

10. Numerical and/or physical models to predict dam breaches are available but modeling could also include substantial uncertainties and subjectivities. When numerical/physical models are not an option, breach parameters should be determined using regression equations. Most regression equations in current use have significant uncertainty in their breach parameter estimations; therefore users need to use good engineering judgment in applying the equations and interpreting the results.

11. Panel 4 focused on dam failures triggered by hydrologic loads, but dam failures can also be the source of flood loading. By design, Panel 4 did not specifically address "sunny-day" failure modes (e.g., seepage erosion / piping), but there is joint USBR/USACE guidance for analyzing these failure modes, which are important potential sources of flood loadings affecting downstream areas. (Similar failure modes can also be triggered by hydrologic events that cause static loading levels that are not commonly experienced) This scenario could be considered a type of "combined event", but it is one that was not specifically covered by Panel 8.

5.6   Abstracts

The workshop organizing committee developed the flooding topics to be discussed and chose the panel co-chairs for each panel topic. The co-chairs then identified potential speakers and discussed them with the organizing committee. Following agreement, the co-chairs sent out invitations to the presenters requesting presentation titles and abstracts for documentation in the workshop program. The speakers and panelists are identified in the workshop agenda (please see Appendix A). The following four abstracts document these presentations and, in some cases, reflect the discussions during the panel session.

5.6.1 Risk-informed Approach to Flood-induced Dam and Levee Failures

David S. Bowles[1,2]

[1]RAC Engineers and Economists, Providence, Utah;
[2]Department of Civil and Environmental Engineering, Utah State University, Logan, Utah.

There is a growing use of probabilistic risk assessment as a supplement to traditional engineering approaches in dam safety decisionmaking in the United States and overseas. This is referred to as a risk-informed approach in which decisionmaking uses information obtained from a risk assessment along with other factors, including good engineering practice and societal concerns. Thus a risk assessment is not the sole basis for a decision, but rather it provides a systematic way of understanding dam failure risk, including the potential consequences and associated uncertainties. Dam safety risk assessments are used for informing decisions about the extent and type of risk reduction and the urgency, priority and phasing of risk-reduction measures.

This presentation will commence with a brief overview of the ownership and regulation of dams in the United States. It will then provide an overview of the major steps involved in performing a risk assessment for an individual dam. Emphasis will be given to flood-related failure modes but reference will also be made to failure modes related to other external and internal hazards and human reliability. The requirements for probabilistic flood hazard information will be summarized, including the effects of spillway gate reliability. Approaches to evaluating the adequacy of dam safety, the tolerability of estimates risks, and using risk analysis outcomes to support dam safety decision making will be provided. An example will be provided for an existing dam and for selecting between risk-reduction alternatives. The paper will close with a summary of some types of information from probabilistic dam and levee flood analysis that might be of value for assessing external hazards for nuclear plants, and some limitations and research needs related to probabilistic risk analysis for flood-induced dam and levee failure.

References

Applications of Dam Safety Risk Assessment
Bowles, D.S., et al, "A Risk-based re-evaluation of reservoir operating restrictions to reduce the risk of failure from earthquake and piping," *2010 ANCOLD Conference on Dams,* Australian National Committee on Large Dams, Hobart, Australia.

Bowles, D.S., and J. Hedien, "Using a Risk-Informed Approach to Find a Solution for Spillway Deficiencies with Upstream and Downstream Tradeoffs," *HydroVision*, July 2012.

Dam Breach Modeling
Wang, Z., and D.S. Bowles, "A Numerical Method for Simulating One-Dimensional Headcut Migration and Overtopping Breaching in Cohesive and Zoned Embankments," *Water Resources Research* 43(5), W05411, 2007.

Chauhan, S.S., D.S. Bowles, and L.R. Anderson, "Do Current Breach Parameter Estimation Techniques Provide Reasonable Estimates for Use in Breach Modeling?", *Proceedings of Dam Safety 2004, ASDSO 2004 Annual Conference,* Association of State Dam Safety Officials, Phoenix, AZ, September 2004.

Wang, Z., and D.S. Bowles, "Overtopping Breaches for a Long Dam Estimated Using a Three-Dimensional Model," *Proceedings of the 2006 USSD Annual Lecture*, United States Society on Dams, San Antonio, TX, June 2006. (For PDF copy of poster, click here.)

Wang, Z., and D.S. Bowles, "Three-Dimensional Non-Cohesive Earthern Dam Breach Model. Part 1. Theory and Methodology," *Advances in Water Resources* 29(10):1528–1545, 2006.

Wang, Z., and D.S. Bowles, "Three-Dimensional Non-Cohesive Earthern Dam Breach Model. Part 2. Validation and applications," *Advances in Water Resources* 29(10):1490–1503, 2006.

Dam Safety Risk Assessment and Management
Bowles, D.S., "Summary of USSD Emerging Issues White Paper on Dam Safety Risk Assessment: What Is It? Who's Using it and Why? Where Should We Be Going With It?", Invited Plenary Session Presentation and Paper in *Proceedings of the 2003 USSD Annual Lecture*, United States Society on Dams, Charleston, SC, April 2003.

Extreme Flood Estimation
Swain, R., D.S. Bowles, and D. Ostenaa, "A Framework for Characterization of Extreme Floods for Dam Safety Risk Assessments," *Proceedings of the 1998 USCOLD Annual Lecture*, United States Committee on Large Dams, Buffalo, NY, August 1998.

Nathan, R.J., and D.S. Bowles, "A Probability-Neutral Approach to the Estimation of Design Snowmelt Floods," Referred paper in *Proceedings of the International Hydrology and Water Resources Symposium*, Auckland, New Zealand, November 1997.

Life-Loss Estimation and Evacuation Modeling
Needham, J.T., Y. Seda-Sanabria and D.S. Bowles, "Consequence Estimation for Critical Infrastructure Risk Management," *Proceedings of the 2010 USSD Annual Lecture*, United States Society on Dams, Sacramento, CA. April 2010.

McClelland, D.M., and D.S. Bowles, "Life-loss estimation: What can we learn from case histories?", *ANCOLD Bulletin* 113:75–91, Australian National Committee on Large Dams, Hobart, Australia, December 1999.

Aboelata, M.A., and D.S. Bowles, "LIFESim: A Model for Estimating Dam Failure Life Loss," Report to the Institute for Water Resources of the U.S. Army Corps of Engineers and Australian National Committee on Large Dams, Utah State University, Logan, UT. 2005.

Aboelata, M.A., and D.S. Bowles, "LIFESim: A Model for Flood Life-Loss Estimation," *Proceedings of the Dam Safety 2008 Conference*, Association of State Dam Safety Officials, Indian Wells, CA, September 2008.

Aboelata, M.A., D.S. Bowles, and A. Chen, "Transportation model for evacuation in estimating dam failure life loss," *ANCOLD Bulletin* 128, Australian Committee on Large Dams, Hobart, Australia, 2005.

Portfolio Risk Assessment and Management
Bowles, D.S., "From Portfolio Risk Assessment to Portfolio Risk Management," *ANCOLD Bulletin* 137:13–32, Australian Committee on Large Dams, Hobart, Australia, 2008.

Bowles, D.S., et al., "Portfolio risk assessment of SA water's large dams," *ANCOLD Bulletin* 112:27–39, August 1999.

Risk Analysis and Probability Estimation
Srivastava, A., D.S. Bowles and S.S. Chauhan, "Improvements to DAMRAE: A Tool for Dam Safety Risk Analysis Modelling," *Proceedings of the ANCOLD Conference on Dams*, Australian Committee on Large Dams, Adelaide, Australia, November 2009.

Risk Analysis—Long Dams
Bowles, D.S., et al., "Baseline Risk Assessment for Herbert Hoover Dike," *ANCOLD Conference on Dams*, Australian Committee on Large Dams, Perth, Australia, October 2012.

Spillway Gate Reliability
Lewin, J., G. Ballard, and D.S. Bowles, "Gate Reliability: An Important Part of Dam Safety," Presented at the *2003 USSD Annual Lecture*, United States Society on Dams, Charleston, SC, April 2003.

Barker, M., B. Vivian, and D.S. Bowles, "Gate Reliability Assessment for a Spillway Upgrade Design in Queensland, Australia," invited paper at the "Workshop on Spillway Reliability," *2005 USSD Annual Conference*, United States Society on Dams, San Antonio, TX, 2005.

Tolerable Risk Evaluation and Decisionmaking
Marsden, J., et al., "Dam safety, economic regulation and society's need to prioritise health and safety expenditures," *Proceedings of the NZSOLD/ANCOLD Workshop on Promoting and Ensuring the Culture of Dam Safety*, New Zealand Society of Large Dams and Australian National Committeee of Large Dams, Queenstown, New Zealand, November 2007.

Munger, D.F., et al., "Interim Tolerable Risk Guidelines for U.S. Army Corps of Engineers Dams," *Proceedings of the 2009 USSD Annual Lecture*, United States Society on Dams, Nashville, TN, April 2009.

Bowles, D.S., "Tolerable Risk for Dams: How Safe is Safe Enough?", *Proceedings of the 2007 USSD Annual Lecture*, United States Society on Dams, Philadelphia, PA, March 2007.

Uncertainty Analysis
Srivastava, A., D.S. Bowles and S.S. Chauhan, "DAMRAE-U: A Tool for Including Uncertainty in Dam Safety Risk Assessment," *Proceedings of the Dam Safety 2012 Conference*, Association of State Dam Safety Officials, Denver, CO, September 17, 2012.

Bowles, D.S., et al., "A Structured Approach to Incorporating Uncertainty into a Dam Safety Risk Assessment," *Proceedings of the 2009 USSD Annual Lecture*, United States Society on Dams, Nashville, TN. April 2009.

5.6.2 Dutch Approach to Levees and Flood Risk

Timo Schweckendiek[1,2]

[1]Deltares, Unit Geo-engineering, Delft, Netherlands
[2]Delft University of Technology, Department of Hydraulic Engineering, Delft, Netherlands

Risk-based approaches in flood defense reliability and flood risk management have a long history in the Netherlands, essentially starting right after a coastal flood disaster in 1953, which inundated large parts of the Dutch southwestern delta and caused more than 1800 fatalities and 100.000 inhabitants lost their homes. In the same year, Van Dantzig (1953) came up with an economic optimization of dike crest levels comparing the investment cost in flood protection with the benefits in terms of risk reduction. The same basic concept is still the basis of flood protection standards nowadays.

More recently, roughly in the last two decades, safety assessment and design of flood defenses are moving steadily from semi-probabilistic approaches towards fully probabilistic reliability and risk analysis. Probably the best example for the state of the practice in the Netherlands is the VNK-2 project (also known as FLORIS; see Jongejan et al., 2013). In the VNK-2 project, all Dutch polders (dike ring systems of so-called "primary flood defenses") are analyzed in terms of probability of failure of the defenses (i.e., levees and hydraulic structures), both at the element and system level, as well as in terms of flood risk. For the latter, sets of inundation scenarios are simulated, the outcomes of which in terms of inundation depth and flow velocities serve as input for modeling the expected damage.

The goal of this presentation is to provide an overview of basic concepts and approaches used for analyzing levee reliability and flood risk in the Netherlands. Special attention will be paid to aspects that may be relevant to PRA of single facilities such as power plants. A major issue here can be the definition of critical flood and failure scenarios addressing the conditions which can pose serious hazards to such assets.

Please find below some references you may find useful and do not hesitate to contact me for further information.

References

Deltacommissie (2008), "Working Together with Water - Findings of the Deltacommissie 2008" (see in particular "Summary" and "Conclusions"), The Hague, The Netherlands, http://www.deltacommissie.com/doc/deltareport_full.pdf.

Faber, M.H. et al. (2007), "Principles of risk assessment of engineered systems," *10th International Conference on Applications of Statistics and Probability in Civil Engineering (ICASP10)*, Taylor & Francis, London, UK, http://www.stanford.edu/~bakerjw/Publications/Faber%20et%20al%20(2007)%20risk%20assessment,%20ICASP10.pdf.

Jongejan, R.B., et al. (2013), "The VNK2-project: a fully probabilistic risk analysis for all major levee systems in the Netherlands," *Floods: From Risk to Opportunity* (in press), International Association of Hydrological Sciences Press, Wallingford, UK, http://www.hkvconsultants.de/documenten/The_VNK2_project_a_fully_probabilistic_risk_etc_B_M_FH(2).pdf.

Jonkman, S.N., et al. (2003), "An overview of quantitative risk measures for loss of life and economic damage," *Journal of Hazardous Materials* 99(1):1–30, http://www.empiwifo.uni-freiburg.de/lehre-teaching-1/summer-term-09/seminar-in-risk-management/materials-seminar-in-risk-management/acceptable_risk9_jonkman.pdf.

Rijkswaterstaat (2005), "Flood Risks and Safety in the Netherlands (Floris)," Delft, The Netherlands, http://ec.europa.eu/ourcoast/download.cfm?fileID=1058.

Schweckendiek, T., et al. (2012), "Target Reliabilities and Partial Factors for Flood Defenses in the Netherlands," *Advances in Soil Mechanics and Geotechnical Engineering, Volume 1: Modern Geotechnical Design Codes of Practice* (in press), Taylor & Francis, London, UK, http://www.academia.edu/3152333/Target_Reliabilities_and_Partial_Factors_for_Flood_Defences_in_the_Netherlands

Schweckendiek, T., et al. (2008), "River System Behavior Effects on Flood Risk," *Proceedings of ESREL 2008*, Valencia, Spain.

Van Dantzig, D. (1953), "Economic Decision Problems for Flood Prevention," *Econometrica* 24(3):276–287, http://oai.cwi.nl/oai/asset/8233/8233A.pdf.

Van der Most, H., and M. Wehrung (2005), "Dealing with Uncertainty in Flood Risk Assessment of Dike Rings in the Netherlands," *Natural Hazards* 36(1–2):191–206.

van Manen, S.E., and M. Brinkhuis, (2005), "Quantitative flood risk assessment for polders," *Reliability Engineering and System Safety* 90(2–3):229–237.

Voortman, H.G. (2003), "Risk-based design of large scale flood defences," Ph.D. thesis, Delft University of Technology, Delft, The Netherlands.

Vrijling, J.K., et al. (1998), "Acceptable risk as a basis for design," *Reliability Engineering and System Safety* 59(1):141–150, http://www.citg.tudelft.nl/fileadmin/Faculteit/CiTG/Over_de_faculteit/Afdelingen/Afdeling_Waterbouwkunde/sectie_waterbouwkunde/people/personal/gelder/publications/citations/doc/citatie66b.pdf.

Vrijling, J.K., T. Schweckendiek, and W. Kanning (2011), "Safety Standards of Flood Defenses," *Proceedings of the 3rd International Symposium on Geotechnical Safety and Risk (ISGSR 2011)*, Munich, Germany, http://www.academia.edu/2078046/Safety_Standards_of_Flood_Defenses.

5.6.3 Risk-Informed Decision-Making (RIDM) Approach for Inflow Design Flood (IDF) Selection and Accommodation for Dams: A Practical Application Case Study

Jason E. Hedien

MWH Americas, Inc., 175 W. Jackson Blvd., Suite 1900, Chicago, IL, 312-831-3095

Previous dam safety investigations at two hydroelectric facilities located on the same river in the Midwestern United States have identified insufficient spillway capacity to pass the computed probable maximum flood (PMF) without overtopping and failing the earth embankments at each project. This presentation provides a case study of how risk assessment (RA) and risk-informed decision making (RIDM) for hydrologic hazard analysis can be applied to select the inflow design flood (IDF) for a particular dam.

The two hydroelectric facilities considered each consist of a multiple-unit surface powerhouse integrated with the dam structures, earth embankment dams using concrete core walls as the primary water barrier, concrete gravity and buttress structures, and spillways. The two facilities are separated by a distance of approximately 12 river miles and are run-of-river projects. Although located in a rural farmland region of the Midwest, both facilities are classified as high-hazard dams. Each dam impounds a lake that is heavily used for recreation during the summer months. The dams are regulated by the Federal Energy Regulatory Commission (FERC).

Per the FERC Engineering Guidelines for the Evaluation of Hydropower Projects, if a dam cannot safely pass the PMF, the dam should be capable of passing the IDF for the dam. In evaluating the options for resolution of the spillway-capacity deficiency at the two dams, the owner and FERC agreed to investigate the use of RA and RIDM to assist in defining the IDF for the dams as an alternative to more traditional incremental hazard justification approaches for computing spillway design floods. A risk-based approach allowed for consideration of several factors specific to these two dams when determining the IDF. These factors include: Both dams are run-of-river dams and do not have flood storage capacity in their reservoirs, which does not allow for attenuation of the flood hydrograph by storage; the population around the lakes impounded by the dams is significantly greater than the population living along the river below the dams, which means that there could be more significant consequences to rises in water levels upstream of the dams prior to dam failure than to rises in water levels downstream of the dams caused by dam failure; the population of the area is significantly greater during the summer months than the winter months, which means that seasonal population differences have an impact on the potential for life loss caused by dam failure; and the winter flood hydrographs are significantly greater in magnitude and rise more quickly than the summer flood hydrographs because of rain on snow causing the snow to melt, along with (a) the lack of vegetative ground cover and (b) frozen, wet ground having relatively smaller surface roughness and less infiltration loss to both hydraulically and hydrologically attenuate runoff towards the river. The RA also allowed for the examination of the effect that failure or non-failure of individual project elements would have on consequences.

The first step in development of the RA for determining the IDF consisted of conducting Potential Failure Modes Analyses (PFMAs) in accordance with FERC guidelines. PFMAs consist of two workshops to identify and evaluate potential failure modes (PFMs) for a given dam, with participants from FERC, the owner, consultants, and specialists. For the dams under investigation, the PFMA workshop was followed by a RA workshop involving the same participants. The PFMA identified a number of credible PFMs that the RA workshop participants agreed should be carried forward into the RA process. These PFMs included: overtopping of the earth embankment at various short and high sections of both dams resulting in failure of the dam; overtopping of a low abutment area at one dam resulting in failure of the right abutment of the dam; and, overtopping of a concrete buttress section at one of the dams resulting in sliding failure of that structure. Each of the PFMs identified would be represented in the RA event tree established for each of the dams.

Event trees representing the sequence of events considered to have the potential to lead to breaching of each dam and uncontrolled release of the reservoir were developed for each dam. Development of the event trees for the dams therefore involved incorporating all PFMs and their interrelationships into appropriate positions, or levels, in the event trees. For all PFMs identified in the PFMA, the initiating event is a given flood inflow into each of the reservoirs impounded by the dams. Thus, the first level in the event tree consisted of the range of possible flood inflows to each reservoir up to and including the inflow associated with the PMF. Because of the significant difference in their magnitude and characteristics, warm and cold season floods were considered separately using the same event tree with the resulting failure probabilities being added to estimate the total failure probability.

Subsequent levels in the event trees included the sequence of events leading to potential breaching of the dam and uncontrolled release of the reservoir. These events, or levels, were divided into those that result in a potential rise in the reservoir impounded by each dam and those associated with the dam failure or breaching process. Events resulting in a potential rise in the reservoir impounded by each dam include the flood inflow itself, failure of spillway gates to open to pass the flood inflow downstream, and potential blockage of spillway gates by debris that hinders passage of flood flow downstream. Events associated in dam failure or breaching included overtopping and initiation of erosion of the embankments of each dam, toppling or failure of core wall segments caused by erosion of the embankments, the initial width of the breach when a core wall segment fails, and sliding of the buttress section at one of the dams.

In the RA each level of the event tree developed for each dam was assigned a conditional probability of occurrence, or system response probability (SRP). The SRP for each event to include in the event trees was estimated during the RA workshop using information and reports available for each dam, historical performance of comparable dams, experience, and engineering judgment. The probability of failure, or the potential for overtopping and breaching of the dam leading to uncontrolled release of the reservoir, was computed by following all of the possible combinations of events and probabilities in the event tree for each dam incorporating all PFMs. The event trees were applied using the complete range of warm and cold season floods to obtain an estimate of the total annual failure probability (AFP) for each dam considering both types of floods as initiating events.

Consequences evaluated for dam breaches and uncontrolled release of the reservoir impounded by each dam included estimated life loss and representative property damage as a count of affected structures. Economic consequences associated with monetary estimates of property damage and other types of economic losses were not estimated. Estimates of life loss and property damage were computed based on the results of breach-inundation model runs using a set of dam-breach and reservoir-release events that cover the full range of warm- and cold-season flood magnitudes for each dam. The number of structures affected by the range of flood events resulting from breaching of the dams was estimated using a database of structures along the river developed from evaluating aerial surveys and photos.

Life loss was estimated using the LIFESim methodology (Aboelata and Bowles, 2005), a modular life-loss modeling approach that considers the effectiveness of warning, evacuation, and rescue; the submergence effects of the flooding and the degree of shelter offered by structures; and the potential for survival or life loss in various flood lethality – shelter environments.

For this RA, the probability of dam failure and associated consequences were compared with tolerable risk guidelines used by the U.S. Department of the Interior's Bureau of Reclamation (Reclamation, 2003) and the Australian National Committee on Large Dams (ANCOLD, 2003) to assess the significance of risks for the baseline risk for the existing dams and various risk-reduction alternatives considered. The ANCOLD and Reclamation guidelines provided a sampling of risk guidelines currently in wide use and applied to dams and reservoirs.

The baseline risk for the dams in their existing conditions was evaluated first. This provided a baseline for evaluating the risks associated with the dams in their current state and the potential benefits/risks and strength of justification provided by structural or non-structural risk-reduction measures considered for the dams. Event tree risk models were developed for the dams in their existing conditions and used to estimate the probabilities and consequences of each of the identified PFMs. The baseline RA results indicated that the dams in their existing condition did not meet either the Reclamation tolerable risk guideline for the AFP or the ANCOLD guideline for the annual life loss (ALL). Based on these results, the owner and FERC determined that risk-reduction alternatives should be investigated.

Investigating the effectiveness of various structural and non-structural risk-reduction measures to accommodate the selected IDFs was carried out as part of the risk-reduction assessment (RRA) for the dams. The risk-reduction assessment (RRA) focused on evaluating the effects of various risk-reduction measures on the probability of dam failure and associated consequences, including both downstream and upstream consequences. The RRA process included the following steps: identifying potential risk-reduction alternatives for consideration in the RRA at each dam; modifying the event trees in the RA model to incorporate risk-reduction alternatives considered for the RRA; developing new peak reservoir pool elevation / annual exceedance probability (AEP) relationships for each risk-reduction alternative; evaluating the effect of each risk-reduction alternative on AFP, number of affected structures, and life-loss estimates relative to those resulting from the baseline risk and tolerable risk guidelines developed by the Reclamation and ANCOLD; and selecting the preferred risk-reduction alternative for further design development with the goal of achieving the most reasonable combination of risk reduction and cost effectiveness with respect to capital cost and ongoing operation and maintenance costs.

The evaluation of a number of risk-reduction alternatives identified for the two dams concluded that adding spillway capacity to each dam by modifying existing structures at each dam to pass additional flows downstream before overtopping failure would meet all tolerable risk guidelines used for the RA and would lower risk to "as low as reasonably possible" (ALARP). In subsequent reviews of the RA results and meetings between FERC and the owner, the additional spillway capacity to be added to each dam was increased somewhat so that the computed warm season PMF could be safely passed during the summer months, when the population that could be impacted by a failure of one of the dams is the greatest. Implementation of these spillway capacity additions was accepted by FERC while satisfying the tolerable risk guidelines cited above and evaluated in the RA. Design and construction of the spillway capacity additions at both dams are currently underway.

The process described in the preceding paragraphs and presented at this workshop provides a case study of how RA and RIDM can be applied to select and accommodate the IDF for a particular dam. In this case, application of RA and RIDM led to the identification of an IDF and spillway capacity improvements that not only are accepted by FERC for selection and accommodation of IDFs, but also reduce computed risks to meet tolerable risk guidelines in use today.

<u>References</u>

Aboelata, M.A., and D.S. Bowles (2005), "LIFESim: A Model for Estimating Dam Failure Life Loss," (Preliminary Draft Report to the Institute for Water Resources of the U.S. Army Corps of Engineers and Australian National Committee on Large Dams), Utah State University, Logan, UT, http://uwrl.usu.edu/people/faculty/DSB/lifesim.pdf.

Australian National Committee on Large Dams (ANCOLD, 2003), "Guidelines on Risk Assessment," Sydney, Australia.

Federal Energy Regulatory Commission (FERC, various years), "Engineering Guidelines for the Evaluation of Hydropower Projects," Washington, DC, http://www.ferc.gov/industries/hydropower/safety/guidelines/eng-guide.asp.

Bureau of Reclamation ("Reclamation," 2003), "Guidelines for Achieving Public Protection in Dam Safety Decisionmaking," Denver, CO, http://www.usbr.gov/ssle/damsafety/Risk/ppg2003.pdf.

5.7.4 USACE Risk Informed Decision Framework for Dam and Levee Safety

David A. Margo, P.E.[1]

[1]U.S. Army Corps of Engineers, Pittsburgh, PA 15222

USACE has transitioned from a solely deterministic standards approach for its dam and levee safety programs to a portfolio risk-management approach. Decisions are now risk-informed with due consideration for all available information. The available evidence is synthesized into estimates of the probability and severity of consequences resulting from breach or malfunction of the dam or levee system. The standards approach also continues to be implemented as an essential input to decisionmaking. Risk-management decisions are made based on the priority and urgency within the portfolio and the cost-effectiveness of risk-reduction options. A tolerable risk framework provides the basis for consistent decisionmaking.

Risk-informed decisions are based on qualitative and quantitative evidence about inundation risks, the degree of uncertainty, the source of the inundation risk, and the efficiency and effectiveness of options to reduce and manage the inundation risk. The source of inundation risk is characterized by the following four scenarios: 1) breach prior to overtopping, 2) overtopping with a breach, 3) malfunction of system components resulting from a variety of causes including human error, or 4) overtopping without breach or non-breach risk. Each of these inundation scenarios represents a source of inundation risk to people, economic activity, vulnerable ecosystems, cultural resources, and other valuable resources and activities.

Risk-Analysis Framework

Risk analysis provides a systematic approach to decision making that enhances the credibility and scientific basis of USACE decisions. Evaluating and reducing risk requires a framework that explicitly considers the risk if no action is taken and recognizes the monetary and non-monetary costs and benefits of risk-reduction options. The USACE framework for risk analysis follows the OMB Principles for Risk Assessment, Management, and Communication. Figure 5–1 depicts the risk analysis framework used by USACE. Risk assessment encompasses a variety of quantitative and qualitative techniques. It is a systematic, evidence-based approach for characterizing the nature, likelihood, and magnitude of risks. Risk management includes actions to identify, evaluate, select, and monitor actions to alter the level of risk. The purpose of risk management is to choose and implement technically sound and integrated actions that reduce risks in an efficient and effective manner. Risk communication is an open, two way exchange of information, opinion, and preference regarding risks and risk-reduction actions that leads to a better understanding of the risks and better decisionmaking.

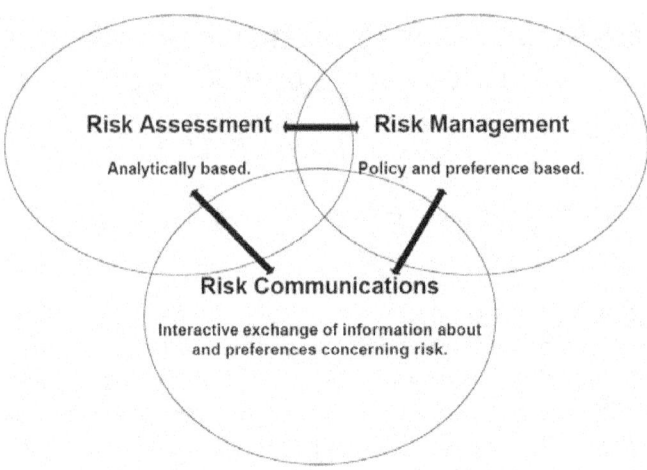

Figure 5–1 Risk analysis framework

Guiding Principles

The following guiding principles, some of which represent a paradigm shift for USACE, have been established for the dam and levee safety programs.

- Life safety will be held paramount in the risk-management decisions made within the USACE dam and levee safety programs. USACE will also consider economic and environmental risks when making decisions.

- The principle of "do no harm" will underpin all actions intended to reduce inundation risk. Applying this principle will ensure that proposed actions do not compromise the overall safety of the facility during or as a result of, implementing a risk-management measure.

- Decisions will be risk-informed (not risk-based). Risk-informed decisions synthesize traditional engineering analyses with risk estimation through the application of experience-based engineering judgment.

- The urgency and priority of actions will be commensurate with the level of inundation risk based on available knowledge.

- USACE will use a portfolio risk-management process to assess and manage the inundation risks associated with the USACE portfolio of dams and levees. This process supports a national-level prioritization to achieve effective and efficient use of available resources.

- The level of effort and scope of risk assessments will be scaled to provide an appropriate level of confidence considering the purpose of the risk-management decision.

- Execution of inspections, instrumentation, monitoring, operations and maintenance, emergency action plans, and other routine activities are an essential part of an effective risk-management strategy.

- Both the USACE and stakeholders share the responsibility and accountability for understanding the inundation risk, formulating and evaluating risk-management options, and selecting and implementing solutions.

Portfolio Risk Management

USACE uses a portfolio risk-management process to systematically assess, characterize, and manage the risks associated with its inventory of dams and levees. A simplified depiction of the process is provided in Figure 5–2. The outer loop of the process includes continuing and recurrent activities of a routine nature. These routine activities are accomplished for all projects regardless of the risk to ensure that risks are being properly and appropriately evaluated, characterized, and managed on a regular basis. When safety issues are identified at a particular facility, a non-routine process is initiated based on the priority and urgency of the issue. Risks are further evaluated in an issue-evaluation study to inform a decision as to whether or not risk-reduction actions may be justified. If justified, then a modification study is conducted to develop and implement specific risk-reduction actions.

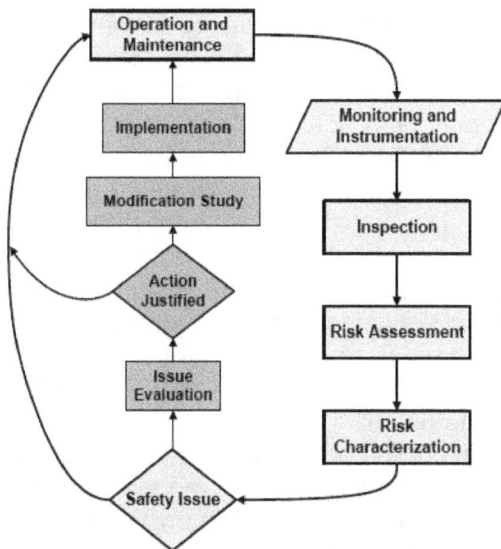

Figure 5–2 Simplified Portfolio Risk Management Process

Estimating Risk

Inundation risks associated with dam and levee systems are estimated by combining the magnitude and likelihood of hazards with the conditional performance of the facility given the hazard and considering the potential consequences that result from each combination of hazard and performance. A comprehensive suite of plausible scenarios and their associated probabilities and consequences are considered, from relatively frequent events to relatively rare events. Evaluation of a limited set of design event scenarios is not sufficient to characterize the risk or support credible decisions. Efforts are generally focused on evaluating those scenarios that have the greatest contribution to the total inundation risk.

Various types of hazard may be plausible (e.g., hydrologic and seismic). These hazards are characterized by all of the relevant parameters (e.g., discharge, stage, duration, velocity, magnitude, and ground acceleration) that may influence the performance or consequences. Depending on the nature of the decision being made, characterization of the seismic hazard might vary from screening-level seismic hazard curves such as those available from the USGS up to a detailed site-specific probabilistic seismic hazard study. Similarly, hydrologic hazard curves may vary in their level of detail from extrapolation of readily available historic information through more detailed studies. Hazard curves typically need to be extrapolated to relatively rare events with annual exceedance probabilities on the order of 10^{-4} to 10^{-6}. This poses a significant challenge because estimates of hydrologic hazard are typically based on a relatively short historic record with various extrapolation techniques. According to USBR (1999), credible extrapolation limits for annual exceedance probability using at-site and regional streamflow data are on the order of 10^{-3}. The limits can be extended to an annual exceedance probability of about 10^{-4} using paleoflood or regional precipitation data. Combining multiple methods using regional data sets may extend the credible extrapolation limit to annual exceedance probabilities on the order of 10^{-5}. Extrapolation beyond these limits based on past experience and historic observations is questionable but often necessary to estimate the inundation risk. The USACE generally considers the probable maximum flood to be an upper bound estimate of the flood hazard for purposes of risk assessment if the estimate is current and represents a "worst reasonable case" scenario.

Identifying, describing, and evaluating the potential performance scenarios that could lead to breach or malfunction of the dam or levee system is one of the most important steps in estimating the inundation risk. The USACE accomplishes this by following a potential failure modes analysis (PFMA) process that typically includes a group of experts in a facilitated team setting. The goal of the PFMA is to identify the important failure modes and to decompose the failure modes into a sequence of steps that are separable, understandable, and for which probabilities can be estimated. The first step of a potential failure mode analysis (PFMA) is to gather the available evidence which should include a site characterization (e.g., the site's geology and geomorphology), documentation of the design basis and construction (e.g., computations, engineering drawings, and photographs), and past performance (e.g., of instrumentation and flood fighting). The second step of a PFMA is to identify candidate failure modes in a brainstorming session. It is important to think beyond deterministic design standard scenarios and consider all of the potential ways in which various conditions and factors can combine in a way that results in breach or malfunction. The brainstorming session often produces an initial list with tens of candidate failure modes. The third step is to narrow this initial candidate list down to those failure modes that are most likely and that have the greatest contribution to the total inundation risk. This can be accomplished by a qualitative assessment of the available evidence considering factors that are both adverse (i.e., breach or malfunction is more likely) and favorable (i.e., breach of malfunction is less likely). This usually results in a list of fewer than ten potential failure modes for which risks need to be estimated. These risk-driver failure modes are then further developed to include a detailed description of the initiating event, the sequence of events that lead to breach or malfunction, the characteristics of the breach (e.g., size and time to develop), and the resulting consequences. The descriptions are then carried over to the risk-estimation process and are used as the basis for developing event trees that depict each event in the failure mode sequence. Probability estimates for each branch of the event tree are typically made using an informal expert elicitation process that considers all of the available evidence (e.g., engineering analysis, past performance, experience with similar projects, judgment).

Inundation scenarios identified by the potential failure mode analysis are used to inform the estimation of consequences. Consequence scenarios are developed to encompass the full range of hazards from relatively frequent events through rare events, variations in breach parameters, and variation in exposure conditions (e.g., daytime vs. nighttime, weekday vs. weekend, and summer vs. winter). An estimate of the magnitude (e.g., depth and velocity) and timing (e.g., arrival time and rate of rise) of inundation is made using hydraulic models and other available information. An initial distribution of people, property, and other assets within the inundated area is made using available datasets such as FEMA HAZUS® or data from site-specific studies. People and assets are then redistributed based on the timing and effectiveness of evacuations relative to the arrival of the flood. Relevant inundation characteristics (e.g., depth, velocity, and rate of rise) are then applied to the remaining people and assets to estimate the loss of life, economic damage, and other impacts.

The hazard, performance, and consequence components of risk are combined across the full range of scenarios either quantitatively or qualitatively to obtain a probability distribution of consequences, usually portrayed as a complementary cumulative distribution or F-N diagram. Average annual risks are also portrayed as the annual probability of inundation versus consequences or f-N diagram.

Uncertainty

Explicit recognition of uncertainties is an essential part of any risk-informed decision. A risk-informed decision to reduce and manage inundation risk is one that considers all available information and the confidence in our knowledge and understanding of the risks and the risk drivers. An important factor to consider is the "known known," which are those factors that are reasonably understood and for which risks and uncertainties can be explicitly estimated. Another important factor to consider is the "known unknown," which are those factors that are known to exist but which may not be well understood. Quantification of the risks and uncertainties for these factors may be difficult due to limited knowledge or lack of available quantitative methods. The final important factor to consider is the "unknown unknown," which are those factors that may not be known or understood. Quantification of the risks and uncertainties for these factors may be impossible because of lack of knowledge and experience.

Uncertainties can be explicitly quantified or qualitatively described for a risk assessment, but it should be recognized that all uncertainties may not be fully identified or captured in the quantitative analysis. Sensitivity analysis is another suggested technique to understand the scenarios, parameters, and assumptions that have the greatest influence on the risk estimate. In addition to quantitative analysis, it is also important to discuss uncertainties within the overall context of the risk assessment by characterizing the state of knowledge using qualitative descriptions. Some example qualitative statements about uncertainty might include: 1) "These are the things about which we have a high degree of confidence," 2) "These are the things that we think are likely, but not proven," 3) "These are the things we think are unlikely, but still might be possible," and 4) "These are the actions we are taking to improve our understanding and reduce the uncertainty." Investments to reduce uncertainty should be made when its impact on the decision is likely to be significant. Decisions on how to deal with uncertainties in the risk estimate may be influenced by agency preferences between a risk-neutral (e.g., best-estimate or most likely outcome) and risk-averse (e.g., reasonable worst-case scenario) approach.

Portraying Risk

Inundation risk associated with a dam or levee may arise from one or more of the inundation scenarios depicted in Figure 5–3. USACE attributes inundation risk to each of the contributing inundation scenarios to provide information on the source of the risk (e.g., performance of the levee, resiliency to overtopping, and magnitude of consequences) and guide the identification and selection of appropriate risk-management options (e.g., strengthen the levee or reduce consequences).

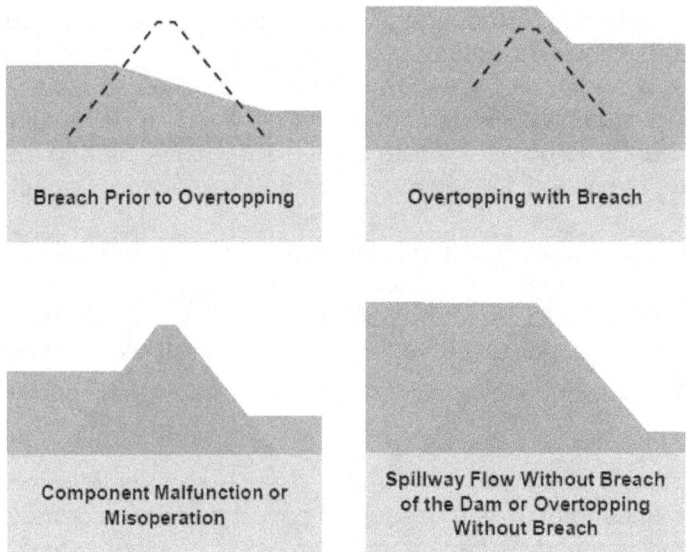

Figure 5–3 Inundation scenarios

The Dam Safety Action Classification (DSAC) and Levee Safety Action Classification (LSAC) are used to guide actions and decisions within the portfolio risk-management process. The classification system provides a consistent and systematic set of guidelines for characterizing the urgency and priority of dam and levee safety actions within the USACE portfolio. Classifications are risk-informed based on a combination of the probability of inundation and the severity of the resulting consequences. Inundation scenarios directly related to performance of the dam or levee (i.e., breach prior to overtopping, overtopping with breach, and component malfunction or misoperation) are explicitly considered when making classification decisions. A classification is made based on available information which typically includes a risk assessment and a technical review by senior dam and levee safety professionals. An initial classification is made based on a screening-level risk assessment and other available information. Over the time, the classification can change based on new information, modifications to the dam or levee system, or changes in the state of engineering practice. The safety action classification framework is displayed in Table 5–1.

Table 5–1 Safety Action Classification Framework

Safety Action Classification Framework (DSAC for Dams and LSAC for Levees)		
Urgency	**Actions**	**Characteristics**
Very High (1)	Actions are recommended for each class commensurate with the level of inundation risk	Risks for each class are characterized by the likelihood of inundation resulting from a breach and/or component malfunction in combination with loss of life and economic or environmental consequences
High (2)		
Moderate (3)		
Low (4)		
Normal (5)		

The average annual inundation risk is portrayed as the annual probability of inundation versus consequences, commonly referred to as an f-N plot, for both quantitative and qualitative risk assessments. From this plot, the average annual inundation risk is obtained as the product of the annual probability of inundation and the consequences. As such, diagonal lines on the plot represent equal levels of annual inundation risk. An example plot for a qualitative risk assessment is provided in Figure 5–4. An example f-N plot for a quantitative risk assessment is provided in Figure 5–5.

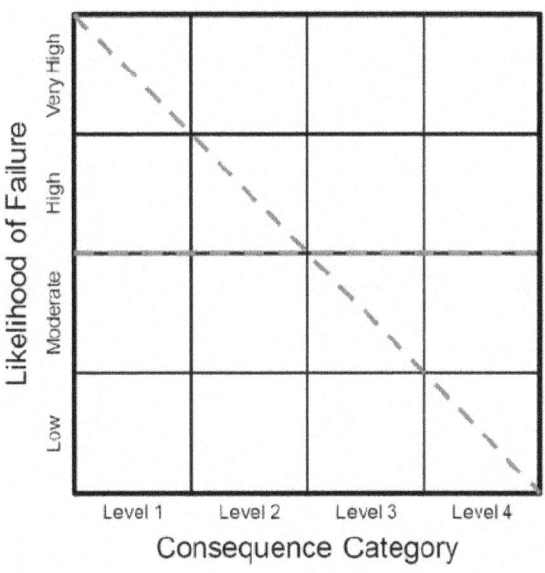

Figure 5–4 Qualitative risk matrix for dams

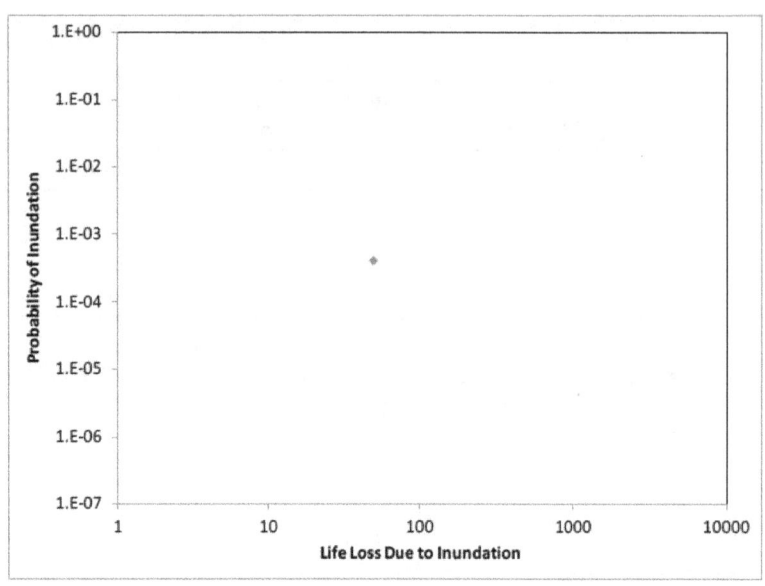

Figure 5–5 Quantitative portrayal of annual risk

A probability distribution of consequences, commonly referred to as an F-N plot, is portrayed as the annual exceedance probability of life loss caused by inundation versus the consequences resulting from inundation. From this plot, the average annual inundation risk is obtained as the area under the relationship. An example plot for a quantitative risk assessment is provided in Figure 5–6.

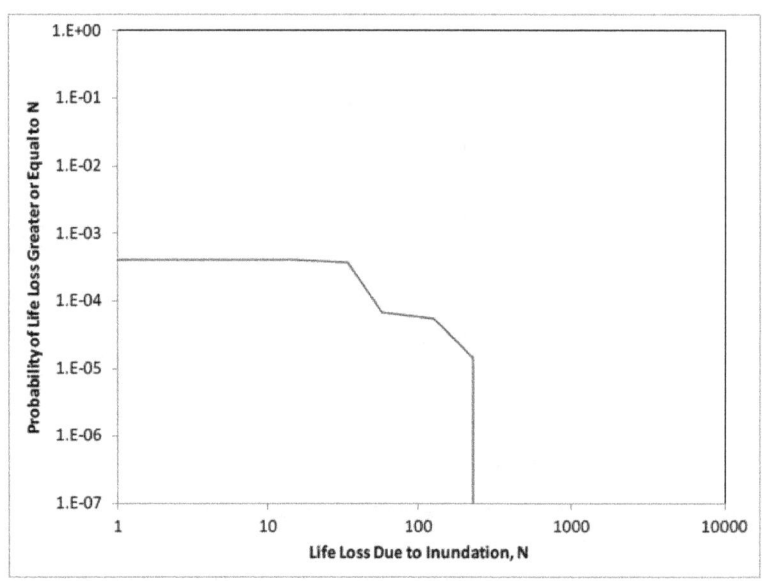

Figure 5–6 Quantitative portrayal of the probability distribution of consequences

Tolerability-of-Risk Framework

The USACE uses a tolerability-of-risk framework, originally developed in the United Kingdom (HSE 2001), to guide decisionmaking within the dam and levee safety programs. Risk may be considered tolerable when each of the following four conditions is satisfied:

- Society is willing to live with the risk to secure benefits.
- Society does not regard the risk as negligible.
- The owner is appropriately assessing and managing the risk.
- The owner reduces the risk when practicable.

The intent of the USACE safety programs is to ensure that inundation risks are managed to achieve and maintain a level that is tolerable in exchange for the benefits afforded by the facility. Tolerability of risk is judged based on the magnitude of both individual and societal risks along with the cost-effectiveness of risk-reduction options, current recognized best practices, and societal concerns.

Individual risk is the increment of risk imposed on a particular individual or group by the existence of a facility. This increment of risk is in addition to the background risk to life, which the person would live with on a daily basis if the facility did not exist. For the USACE dam and levee safety programs, individual inundation risk is defined as the risk of a fatality to the most at-risk individual. It varies by location and factors that affect exposure and vulnerability of the individuals. The spatial distribution of individual inundation risk can support formulation of options to reduce inundation risk as well as provide a framework for comparing the risk-reduction performance of various options. Societal risk is the risk of widespread or large-scale loss caused by inundation, the implication being that the consequences would provoke a socio/political response, and/or that the risk provokes public reaction and is effectively regulated by society as a whole through its political processes and regulatory mechanisms. Thus, while providing benefits to society by enabling compatible use of the floodplain, the presence of a dam or levee system also represents a risk to those in the floodplain wherein there may be widespread, multiple life loss should the floodplain become inundated. In general, society is believed to be more averse to risks if multiple fatalities were to occur from a single event and hence impact society as a whole. In contrast, society tends to be less averse to risks that result from many events resulting in only one or two fatalities, even if the total loss represented by the sum of fatalities from all of the small-loss accidents is larger than that from the single large-loss accident.

Decisionmaking

A risk-informed decision is based on qualitative and quantitative evidence about inundation risk, which of the inundation scenarios is creating that risk, and the effectiveness of options to reduce and manage this inundation risk. Explicit recognition of uncertainties is a part of a risk-informed decision, because such uncertainties can influence the selection of risk-management options. A risk-informed decision to reduce and manage flood inundation risk is one that considers all options to reduce and manage inundation risk, the available resources to implement risk-reduction and -management measures, the efficiency and effectiveness of each option, and the specific recommendation to implement the particular option.

The nation has constructed many thousands of dams and levees. Today's challenge is to create a risk-informed decisionmaking approach that will lead to a better understanding of the inundation risk associated with these facilities. Risk-informed decisionmaking illuminates the choices that must be made to address the issues of tolerable inundation risk. Managing inundation risk is a shared responsibility of individuals and governments, each with their own objectives, authorities, and limitations. Different circumstances, different perceptions of the benefits of risk-taking, and different attitudes regarding risk may lead to different risk-reduction options and decisions by different entities. There is no best level of inundation risk reduction and no optimal combination or sequence for risk-reduction actions. Inundation risk has many dynamic aspects that include actions by USACE, local sponsors, various stakeholders, and the public. The intended risk-management outcomes are influenced in varying ways by action or inaction by each of these entities. Figure 5–7 illustrates how risks can change over time as a result of various actions or inaction.

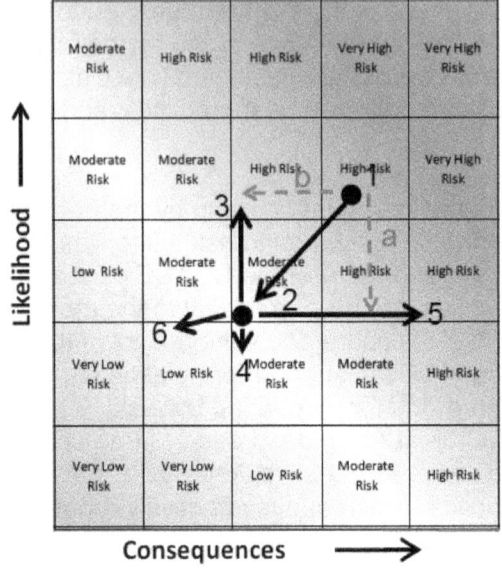

1. Current risk
2. Risk after implementation of risk management options
   a. *Actions that reduce likelihood*
   b. *Actions that reduce consequences*
3. Aging and wear and tear.
4. Proper maintenance, repairs, and operations
5. Floodplain development
6. Continued risk management activities

Figure 5–7 Trends in inundation risk over time

References

Bureau of Reclamation ("Reclamation," 1999), "A Framework for Characterizing Extreme Floods for Dam Safety Risk Assessment," prepared by Utah State University and the Bureau of Reclamation, Denver, CO.

U.S. Army Corps of Engineers (2011), "Safety of Dams – Policy and Procedures," Engineer Regulation 1110-2-1156, Washington, DC, http://publications.usace.army.mil/publications/eng-regs/ER_1110-2-1156/ER_1110-2-1156.pdf.

Health & Safety Executive (HSE, 2001), "Reducing Risks, Protecting People: HSE's Decision-making Process," Risk Assessment Policy Unit, HSE Books, Her Majesty's Stationery Office, Norwich, England. http://www.hse.gov.uk/risk/theory/r2p2.pdf.

PANEL 5

TSUNAMI FLOODING

Co-Chairs:
Eric Geist, USGS
Henry Jones, NRC

Technical Reporters:
Mark McBride and Randy Fedors, NRC

6.   Tsunami Flooding

Eric Geist[1], Henry Jones[2], Mark McBride[2], and Randy Fedors[4]

[1]U.S. Geological Survey, Menlo Park, CA
[2]Office of New Reactors, U.S. NRC, Rockville, MD
[4]Office of Nuclear Material Safety and Safeguards, U.S. NRC, Rockville, MD

6.1   Motivation

Panel 5 focused on tsunami flooding with an emphasis on Probabilistic Tsunami Hazard Analysis (PTHA) as derived from its counterpart, Probabilistic Seismic Hazard Analysis (PSHA) that determines seismic ground-motion hazards. The Panel reviewed current practices in PTHA and determined the viability of extending the analysis to extreme design probabilities (i.e., 10^{-4} to 10^{-6}). In addition to earthquake sources for tsunamis, PTHA for extreme events necessitates the inclusion of tsunamis generated by submarine landslides, and treatment of the large attendant uncertainty in source characterization and recurrence rates. Tsunamis can be caused by local and distant earthquakes, landslides, volcanism, and asteroid/meteorite impacts. Coastal flooding caused by storm surges and seiches is covered in Panel 7. Tsunamis directly tied to earthquakes, the similarities with (and path forward offered by) the PSHA approach for PTHA, and especially submarine landslide tsunamis were a particular focus of Panel 5.

6.2   Background

Past formulations and applications of PTHA have primarily included only earthquake sources. For these studies, there is a sufficient catalog of sources along most fault zones so that critical parameters of PTHA can be estimated with quantifiable uncertainty. These parameters include distribution of tsunami source sizes (i.e., earthquake magnitude) and mean recurrence rate. Including submarine landslides in PTHA at extreme design probabilities has only recently been considered. Because of a lack of an instrumental catalog for submarine landslides as for earthquakes, geologic and geophysical methods must be used to determine source sizes and recurrence. Moreover, the hydrodynamics of landslide tsunamis are considerably more complex, necessitating the use of higher-order nonlinear wave equations.

Panel 5, "Tsunami Flooding," addressed advanced methods for PTHA, with an overall emphasis on defining tsunami hazards at extreme design probabilities. The primary differences between PTHA and its predecessor PSHA are the inclusion and characterization of far-field (i.e., transoceanic) and non-seismogenic sources and the use of numerical propagation models in place of ground-motion attenuation relations that are used in PSHA. Much of the work discussed in Panel 5, therefore, relates to proper characterization of tsunami sources and development of robust tsunami models that can accurately simulate tsunami propagation, runup, and inundation.

6.3   Overview of Presentations

There were four presentations in Panel 5 plus an additional short presentation during the discussion period. The first half of the Panel discussed the fundamentals of PTHA as well as recent advances. The presentation and abstract by Thio outlined the basic procedure of PTHA, focusing on how aleatory and epistemic sources of uncertainty are incorporated. LeVeque presented new techniques implemented in a PTHA study for Crescent City, California

(Gonzalez, et al., 2013). These include using pattern-detection methods to incorporate tidal uncertainty in PTHA and using a Karhunen-Loeve expansion to represent slip distributions on non-planar faults. Statistical testing of hypotheses and assumptions inherent to PTHA is described in an abstract by Geist et al. provided with the workshop proceedings. In an abstract by Real and Wilson, application of PTHA to tsunami hazard mapping objectives in California is described.

The second half of the Panel discussed issues specifically related to landslide tsunamis, as they most often dictate the hazard at extreme probabilities, especially along passive margins. The presentation and abstract by ten Brink et al. details procedures for developing landslide size and occurrence distributions using marine geological and geophysical information. Oceanographic and geological conditions (such as long-term sea-level changes and sea-floor physiography/cohesion, respectively) and earthquake-triggering mechanisms present a complex parameter space to determine the probability of landslide failure. For a given set of submarine landslide parameters, Lynett presented hydrodynamic methods to simulate tsunami propagation and inundation, often involving nonlinear and dispersive properties.

During the discussion session, Wei and Titov described how tsunami forecast tools recently developed at NOAA can be an important component of tsunami flooding assessment and of PTHA in particular. In their abstract, they indicate the importance of using propagation and inundation models that have been thoroughly developed and tested to provide the best possible accuracy for hazard assessments.

6.4   Summary of Panel Discussion

The questions posed to the panel during the discussion period were divided among three groups: (1) general PTHA questions; (2) landslide PTHA questions; and (3) questions specific to the implementation of PTHA. The questions were discussed among panel members before the workshop as well as during the panel discussion session.

Group (1) Questions: General PTHA

For group (1), the first question was "What input parameters/uncertainties are important to include in PTHA for extreme tsunamis?". The primary parameters appear to be the source size distribution and recurrence rate and distribution. In addition, source parameters that scale with the size of the generating event are important, as are those parameters that are independent of the source size, such as rise time and rupture velocity. Bottom friction is likely the primary propagation/inundation parameter that dictates the severity of runup.

The second question in group (1) was "What are the appropriate probability distributions (as determined by statistical testing and model selection) that define uncertainty of input parameters for PTHA? What databases exist for earthquakes and landslides to test model distributions and assess uncertainty?" Much research has been done on defining the size and recurrence distributions of tsunami sources. For example, many studies use earthquake catalog data to formally test competing hypotheses, such as the characteristic and Gutenberg-Richter hypotheses for size distributions. Much less is known about how other tsunami parameters are distributed. Some of the tsunami source parameters from the 2011 Tohoku earthquake were very unusual compared to past subduction-zone earthquakes of this size. More work needs to be done in compiling the range of tsunami source parameters from past earthquakes and landslides so that the appropriate distributions can be determined. For many subduction fault zones, the geometry of the region most likely to slip, along with expected tapering and correlation lengths of slip patterns, needs to be better known in order to develop better source parameter distributions for stochastic sampling.

The final question in group (1) was, "How do PTHA techniques currently used in the United States differ from those implemented in Japan, especially with regard to extreme tsunami hazards?". The primary journal papers that outline the Japanese approach to PTHA are Rikitake and Aida (1988) and Annaka et al (2007). The primary differences are that characteristic earthquake distributions and quasiperiodic recurrence are frequently adopted in Japan, whereas the Gutenberg-Richter size distribution has been adopted in some U.S. PTHA studies. In addition, there is a recent focus on landslide sources in the United States for PTHA at extreme probabilities, whereas landslides are most often not considered in Japanese PTHA studies. PTHA studies have also been completed in Australia and New Zealand (e.g., Burbridge and others, 2008).

Group (2) Questions: Landslide PTHA

The first question in group (2) was, "What is the best framework (e.g., logic-tree) for including landslide tsunamis in PTHA, given the inherent uncertainties associated with the phenomenon?" There have been several different approaches to modeling submarine landslides (as well as other types of landslides) that can be included in a PSHA-style logic tree. Because these approaches are all very much in constant development and are judgment-based, it may be more practical to set up an overall probabilistic framework in which different researchers can contribute their specific model results (and provide aleatory uncertainties), which are then combined into a final probabilistic product. PSHA is usually done using some monolithic codes, but at the heart of it are some very compartmentalized processes. There is quite a bit of data on landslide size distributions, which may be enough to get started. However, we don't have a lot of geologic and oceanographic data with which to constrain recurrence rates (and bounds), although probabilistic techniques are being developed where this data does exist. The geometry of the observed slides (and their scars) gives us at least a constraint on the dynamic behavior of the slides and possibly the physical parameters (e.g., rheologic parameters) that are acceptable for dynamic simulations. Specific challenges for including landslides in the PTHA were discussed in a 2011 joint NRC/USGS workshop (Geist and ten Brink, 2012). As with earthquake sources, more work needs to be done in compiling the range of tsunami source parameters from past events so that the appropriate distributions can be determined.

The second question in group (2) was, "How does a landslide composed of multiple failures generate a tsunami? By constructive interference of many small failures, by only failure of a dominant large cohesive block, or by hydraulic jump of thick coalesced debris flows from a large area?". There is mounting evidence from the 1929 Grand Banks landslide that the continental slope in the vicinity of the triggering M7.2 earthquake failed in many small sub-events. The question becomes, how did these small failures result in a damaging tsunami 3 to 4 m high? One hypothesis is that the debris flows from each of the subevents coalesced in a single region and that some dynamic phenomena accentuated the tsunami-generating capacity. In contrast to simple single-event landslide scenarios that are typically used in tsunami modeling, many real-world landslides exhibit complex failure behavior in both space and time, such as was observed from the 1929 Grand Banks landslide. Newly developed debris-flow modeling codes (e.g., extensions of GeoClaw being developed at the USGS Cascades Volcano Observatory) will likely aid in answering this type of question.

Group (3) Questions: PTHA Implementation

The first question in group (3) was, "What are the fundamental needs of improving PTHA (e.g., seafloor mapping/sampling, paleoseismic/paleotsunami deposit analysis, validation of modeled current velocities, etc.), and how should the United States move forward to make these improvements (organizationally and to secure funding)?". As reiterated from previous questions, the primary need in improving PTHA is a comprehensive database of tsunami-generation parameters, particularly those of landslide tsunamis. This database would allow a better determination of parameter uncertainty and appropriate distributions than currently exists.

Other outstanding scientific problems include understanding tsunamis generated by the outer-rise earthquakes (including their source locations and rupturing mechanism). Recent studies of the 2009 Samoa, 2010 Haiti, and 2011 Tohoku tsunamis have indicated the complication caused by composite source mechanisms. Also, how the energy is transferred from seismic activity to the ambient water body is poorly understood. Improvement of the source model may be needed to improve PTHA for numerical model engagement.

In terms of acquiring new data on tsunami sources, with diminishing opportunities for new funding for applied earthquake and tsunami research, there is a need for greater inter-program/multi-agency coordination and pooled funding to sustain current efforts. Improved coordination between (and support for) USGS, NOAA, and state geological surveys is needed to collect offshore information about the location and activity of faults and landslides. For implementation of PTHA at extreme probabilities, the NRC should work more closely with other agencies and programs (USGS, NOAA, FEMA, the National Tsunami Hazard Mitigation Program, and state geological surveys) to explore how to fund and collect information vital to the development of more reliable PTHA.

The second question in group (3) was, "How should the NRC and other organizations work towards verification and consistency of PTHA methods?". One possible verification method is to perform a probabilistic analysis based on empirical data only and compare the results against the PTHA hazard curve derived from computational methods. If the empirical probability estimate is significantly higher than the computational estimate, something is wrong. Empirical estimates will be strongly affected by censoring and catalog completeness/duration. Verification of extreme tsunamis is problematic.

A number of government programs have interests in exploring, or partial mandates to explore, PTHA methods. The National Tsunami Hazard Mitigation Program (NTHMP)—which has representation from all coastal states/territories, the USGS, FEMA, and NOAA—has the platform to develop guidance/work plans for data collection, database development, model validation, etc. Though funding within the program is limited, the NTHMP is exploring how to develop PTHA products in a nationally consistent and cost-effective manner. For example, the NTHMP is funding the State of California to evaluate two PTHA methods in Crescent City and begin PTHA map production for multiple uses (land use, consistent evacuation, building/construction, etc.); the California Geological Survey has developed an expert Work Group to assist in this review. The American Society of Civil Engineering (ASCE) 7 Tsunami Loads and Effects Subcommittee is working to incorporate Tsunami Design Maps for buildings and other structures to be produced by PTHA for U.S. West Coast, Alaska, Hawaii, and the Pacific Islands. It will be carried out through a collaborative effort of ASCE, NOAA/PMEL, the University of Washington, URS Corporation, and the California Geological Survey. The NRC can explore partnering with the NTHMP and the State of California in this effort to help with coordinated background data collection, improving numerical modeling, and PTHA method verification. For example, followup on the insights gained from the Crescent City test and the Work Group's evaluation, which can lead to further improvements in the PTHA models tested, will require continued support.

6.5   Observations and Insights

In relation to other panels describing flood hazards, it was apparent that PTHA is more closely allied to PSHA than probabilistic analysis of other flood hazards. Large tsunamis are thankfully infrequent; however, this results in tsunami catalogs of limited length, and making empirical analyses at extreme probabilities is highly dependent on distribution selection and catalog errors and censoring. In contrast, a computationally based PTHA approach, based on numerical propagation and inundation models, uses earthquake catalogs that contain substantially more data. Much of Panel 5 focused on how submarine landslides can also be incorporated in a computational PTHA. Insights were gained from other Panels that developed PHA for other flood hazards, potentially resulting in improvements to PTHA. In particular, joint probability methods that account for correlation (e.g., Toro et al., 2010) among tsunami parameters has in the past been largely ignored. Although there are likely fewer dependent parameters in PTHA than, for example, storm surge, multivariate analysis should be considered in future PTHA studies.

It is suggested that:

*   PTHA should follow the path of the PSHA process of the past 5 years, both for compilation of appropriates maps and filling of data gaps by convening expert panels

*   Progress should be made towards further development of data to support catalogs or maps of landslide characteristics and probabilities

*   Better incorporation of the physical processes should be made in models for tsunami generation, runup, and inundation

6.6 References

Annaka, T., Satake, K., Sakakiyama, T., Yanagisawa, K., and Shuto, N., 2007, Logic-tree approach for probabilistic tsunami hazard analysis and its applications to the Japanese Coasts," *Pure and Applied Geophysics* 164:577–592.

Burbridge, D., Cummins, P.R., Mleczko, R., and Thio, H.K., 2008, "A probabilistic tsunami hazard assessment for western Australia," *Pure and Applied Geophysics* 165:2059–2088.

Geist, E.L., and ten Brink, U.S., 2012, "NRC/USGS Workshop Report: Landslide Tsunami Probability, Convened August 18-19, 2011 at the USGS Woods Hole Science Center, Woods Hole, MA," U.S. Geological Survey Administrative Report, Reston, VA. Text: http://woodshole.er.usgs.gov/staffpages/utenbrink/my%20publications/ landslide_ WorkshopReport.pdf; Appendices, http://woodshole.er.usgs.gov/staffpages/utenbrink/ my%20publications/landslide_ Workshop_abstracts_presentations.pdf; accessed March 18, 2013.

Gonzalez, F.I., R.J. LeVeque, and L. Adams, 2013, "Probabilistic Tsunami Hazard Assessment (PTHA) for Crescent CIty, CA," Final Report for Phase I, February 2, 2013. Pilot study funded by BakerAECOM. Obtained 03/20/2013 from http://faculty.washington.edu/rjl/pubs/CCptha. Rikitake, T., and Aida, I., 1988, "Tsunami hazard probability in Japan," *Bulletin of the Seismological Society of America* 78:1268–1278.

Toro, G.R., Niedoroda, A.W., Reed, C.W., and Divoky, D., 2010, "Quadrature-based approach for the efficient evaluation of surge hazard," *Ocean Engineering* 37:114–124.

6.7 Abstracts

The workshop organizing committee developed the flooding topics to be discussed and chose the panel co-chairs for each panel topic. The co-chairs then identified potential speakers and discussed them with the organizing committee. Following agreement, the co-chairs sent out invitations to the presenters requesting presentation titles and abstracts for documentation in the workshop program. The speakers and panelists are identified in the workshop agenda (please see Appendix A). The following seven abstracts document these presentations and, in some cases, reflect the discussions during the panel session.

6.7.1 Probabilistic Tsunami Hazard Analysis

Hong Kie Thio

URS Corporation, Los Angeles, CA 90017

The 2004 Sumatra tsunami disaster has accelerated the development of performance-based solutions to the analysis of tsunami hazard and thus the establishment of a probabilistic framework for tsunami hazard analysis. We have developed a method for Probabilistic Tsunami Hazard Analysis (PTHA) that closely follows, where possible, the methodology of Probabilistic Seismic Hazard Analysis (PSHA) using a hybrid approach. Because of the very strong variability of the seafloor depth, which causes very strong distortions of the wavefield and thus the wave amplitudes, it is not practical to develop analogs to the ground motion prediction equations (GMPEs) used in PSHA. Instead, we can use numerical models to directly simulate tsunami wave propagation because the structure of the oceans (i.e., their bathymetry) is generally well-known. In contrast to the deep ocean propagation where the long tsunami waves behave in a linear fashion, the nearshore propagation, inundation, and runup are highly nonlinear and require a far greater computational effort. We have therefore split the PTHA analysis into two steps, which is somewhat similar to the seismic practice where we often compute probabilistic spectra for rock conditions using simple (quasi-)linear relations, and use these as input for nonlinear analysis of the soil response.

In the first step, we use the linear behavior of offshore wave propagation to compute fully probabilistic offshore waveheights using a library of pre-computed tsunami Green's functions for elementary fault elements (usually 50 × 50 km) from all potential sources. This allows us to efficiently perform an integration over a wide range of locations and magnitudes, similar to PSHA, and incorporate both epistemic uncertainties (through the use of logic trees) and aleatory uncertainties using distribution functions. We have developed maps of probabilistic offshore waveheights for the western United States for a range of return period between 72 and 2500 years. These maps are very useful to compare the hazard between different regions and, through disaggregation, identify the dominant sources for that particular region and return period. The disaggregation results are then used in the next step in which we compute inundation scenarios using a small subset of events that are representative of the hazard expressed by the probabilistic waveheights. This process has enabled us to develop probabilistic tsunami inundation maps for the state of California.

Epistemic Uncertainties

An important aspect of the probabilistic approach, especially for longer return periods, is the proper characterization of the uncertainties. Because tsunamis can still be damaging at very large distances, as was demonstrated by the 2004 Sumatra and 2011 Tohoku earthquakes, it is important to include sources from the entire Pacific Basin for a hazard study of the U.S. west coast. Our state of knowledge of these different source zones is however highly variable, and in most cases, the record of observations is rather short and incomplete. We therefore developed a "generic" source characterization based on global scaling relations, plate convergence rates and some assumptions regarding stress-drop and seismogenic thickness. This model can be used in its entirety for subduction zones for which we have little other information, but is also included, as a separate logic tree branch, in cases for which we do have more specific information on the fault. Depending on the quality of the constraints we have for the specific source, such as Cascadia, the relative importance (weight) of the generic model is adjusted. As is the case in PSHA, our aim is to strike a balance between empirical models, which tend to be incomplete in terms of the length of record and may include ambiguous observations, and an imperfect understanding of the behavior and recurrence of large tsunamigenic earthquakes.

Aleatory Uncertainties

We include aleatory variability for the different elements of the hazard analysis, including the source characterization, bearing in mind that the large slip (~50 m) observed for the Tohoku earthquake corresponds approximately to the 2 sigma level for maximum slip in the global scaling relations. It is therefore important to be cautious with the truncation of the aleatory distributions. Whereas GMPEs in seismic hazard analysis automatically include a standard deviation, the numerical approach does not, and we therefore have computed a modeling sigma by comparing modeled tsunami waveheights with observed ones. Finally, we also include tidal uncertainty by convolving the tsunami Green's functions with the local tidal record and developing a distribution function of the maximum waveheights at every step.

Future work

We have established a framework for performing probabilistic tsunami hazard analysis, which is currently focused on earthquake sources but may also include other types of sources such as submarine landslides, for which the range of uncertainties is much larger. Even for the earthquake problem, there are many important issues outstanding, for instance with regard to the very large (M > 9) earthquakes. Accurate characterization of the upper end of the magnitude range of a source is often, perhaps somewhat counterintuitively, not very important in seismic hazard because the ground motions tend to saturate for very large magnitudes. This is however a critical issue for tsunami hazard analysis, and it is very important to understand the limits for these large earthquakes. In many cases we may be able to assign a maximum rupture length, simply by considering the geometry of the subduction zone, but it is important to understand how the slip scales at these very large magnitudes, for which we have very little data. If we assume some maximum stress-drop arguments, it is possible to constrain the slip using the fault width, but its scaling for large magnitude is also poorly constrained. Likewise, it is important to determine how often these events occur, or whether segmented ruptures are the main mode of rupture. It is therefore essential to collect more empirical data, in particular from paleo-tsunami fieldwork, to extend the tsunami record and provide better constraints on or recurrence models.

6.7.2   Recent Advances in PTHA Methodology

Randall J. LeVeque[1], Frank I. Gonzalez[2], Knut Waagan[1], Loyce M. Adams[1], and Guang Lin[3]

[1]University of Washington, Applied Mathematics
[2]University of Washington, Earth and Space Sciences
[3]Pacific Northwest National Laboratory, Computational Sciences and Mathematics

Probabilistic tsunami hazard assessment (PTHA) techniques have been developed by many researchers over the past decade, largely based on the older field of probabilistic seismic hazard assessment (PSHA). For surveys see Geist and Parsons (2006) and Geist et al. (2009).

Recently the PTHA methodology employed in a study of Seaside, OR (Gonzalez et al., 2009) has been updated in connection with a recent investigation of tsunami hazards in Crescent City, CA.

The original Seaside methodology uses the following steps:

1. Determine a finite set of potential tsunamigenic earthquakes together with annual probabilities of occurrence.

2. Simulate the tsunami resulting from each, and the resulting inundation in the target region of interest.

3. Incorporate tidal uncertainty to adjust the probabilities of flooding a given point from each event.

4. At each point on a fine-scale spatial grid covering the region, construct a "hazard curve" giving the annual probability of exceedance as a function of some range of exceedance values.

5. Determine, for example, the 100-year flood depth by interpolating on this curve to find the depth that is exceeded with annual probability 0.01.

6. Combine these values obtained from the hazard curves for all spatial points to produce a map of the 100-year flood, and contours of the depth that is exceeded with probability 0.01.

The updated methodology makes use of the open source GeoClaw software, which uses adaptive mesh refinement (AMR) to efficiently solve the tsunami propagation and inundation problem, allowing the use of large numbers of potential tsunamis and the possibility of sampling from probability density functions of potential earthquakes rather than using a small set of characteristic events. Past work in this direction has often used a small number of stochastic parameters to characterize the earthquake, such as depth and magnitude (see e.g. Geist and Parsons, 2006), which can be very effective for distant sources for which the runup and inundation are relatively insensitive to the details of the slip distribution. Recent work has focused on exploring the use of a Karhunen-Loève expansion to represent possible slip distributions on an earthquake fault for the nearfield case. This is an expansion in terms of eigenfunctions of a presumed covariance matrix for the slip distribution over the fault geometry.

Guidance on the choice of covariance matrix based on the dimensions of the fault plane and earthquake magnitude can be found in work such as that of Mai and Beroza (2002). The coefficients in this expansion specify a particular realization of this potential earthquake. If a probability density function in the high-dimensional space of coefficients can be determined that accurately describes potential earthquakes, then importance sampling and dimension reduction techniques can be used to derive hazard curves from this stochastic description of potential events. Unfortunately, there is a high degree of epistemic uncertainty that makes it difficult to adequately constrain these density functions for regions such as the Cascadia Subduction Zone, and more work is needed in this direction.

The GeoClaw code also allows the direct computation of inundation for different tide stages. This allows tidal uncertainty to be handled more accurately than in the Seaside study, where the approach of Mojfeld et al. (2007) was used. For each tsunami event studied, inundation is typically computed at three tide stages: mean low water (MLW), mean sea level (MSL), and mean higher high water (MHHW). At each spatial point on the regional grid, and for each exceedance level, these three simulations can be used to estimate the tidal stage that would be necessary for inundation that exceeds the given level. The tide record can then be used to compute probabilities that this tide stage or higher will occur when the tsunami arrives. For tsunamis that consist of multiple large waves, the wave pattern unique to the tsunami can be used to improve on these probabilities.

Interpretation of the resulting hazard curves and maps can be improved by considering methods of plotting the data. Traditional flood maps such as those showing contours of a 100-year flood can be supplemented by contour maps of probability for a given exceedance value. These can be important in understanding the sensitivity of results to probability level chosen, and will better reveal the possibility of much more extreme flooding events that may occur with slightly smaller probability.

Another extension is to study flow velocities, momentum, and momentum flux in addition to flow depth. Forces on structures and the destructive capacity of the flow increase with velocity, particularly when there is debris carried in the fluid. Hazard maps that only consider depth of flow may be inadequate for judging the probability of disaster.

References

Gonzalez, F. I., E.L. Geist, B. Jaffe, U. Kanoglu, 2009. "Probabilistic tsunami hazard assessment at Seaside, Oregon, for near- and far-field seismic sources," *Journal of Geophysical Research* 114:C11023, http://profile.usgs.gov/myscience/upload_folder/ci2012Jun22202847426802008JC005132.pdf.

Geist, E.L., and T. Parsons, 2006, "Probabilistic Analysis of Tsunami Hazards," *Natural Hazards* 37:277–314, http://walrus.wr.usgs.gov/reports/reprints/Geist_NH_37.pdf.

Geist, E.L. T. Parsons, U. S. ten Brink, and H. J. Lee, 2009, "Tsunami Probability," E.N. Bernard and A.R. Robinson (eds.), *The Sea* 15:201, Harvard University Press, Cambridge, MA.

Mai, P.M., and G.C. Beroza, 2002, "A spatial random field model to characterize complexity in earthquake slip," *Journal of Geophysical Research* 107(B11):2308, http://www.seg2.ethz.ch/dalguer/courses/earthquake/Mai_and_Beroza2002JGR.pdf.

Mofjeld, H.O., F.I. Gonzalez, V.V. Titov, A.J. Venturato, and J.C. Newman, 2007. Effects of tides on maximum tsunami wave heights: Probability distributions," *Journal of Atmospheric and Oceanic Technology* 24:117–123, http://journals.ametsoc.org/doi/pdf/10.1175/JTECH1955.1.

6.7.3 Geological Perspective on Submarine Landslide Tsunami Probability

Uri S. ten Brink[1], Jason D. Chaytor[1], Daniel S. Brothers[1], and Eric L. Geist[2]

[1]U.S. Geological Survey, Woods Hole, MA 02543
[2]U.S. Geological Survey, Menlo Park, CA 94025

The temporal distribution of earthquakes, other than aftershocks and triggered events, is assumed to follow an exponential distribution that is associated with a stationary Poisson's process (e.g., Parsons, 2002; Coral, 2004; and Kagan, 2010). The spatial distribution of earthquakes is heterogeneous, being concentrated in tectonic plate boundaries and other tectonically active zones, notwithstanding the infrequent intra-plate earthquakes (e.g., Swafford and Stein, 2007).

Are submarine landslide distributions uniform in time and space? Our knowledge of submarine landslide distribution in space is limited by the difficulty of mapping the seafloor at high resolution. Nevertheless, our knowledge improves as more bathymetry and high-resolution seismic data are being collected. For example, the U.S. Atlantic continental margin from North Carolina north is almost completely covered by bathymetric data allowing us to map all but the very smallest (< 1 km^2) landslide scars. Inferring the temporal distribution of submarine landslide distribution is more challenging because of our inability to detect slope failures in real time, the considerable work involved in robustly dating landslide features (e.g., Haflidason et al., 2005), and sometimes because of the lack of, or technical limitations of recovering, dateable material. Moreover, landslides, being a destructive process, often erase the record of previous slides.

It is therefore necessary to use indirect arguments to constrain landslide distributions. One such argument is the ergodic hypothesis in which the age distribution of landslides around the globe can yield the rate of landslide occurrence at a particular location. Caution should be taken in using this argument because of the potential differences between seismically active and "passive" (i.e., non-active) margins, and between margins that were influenced by glaciers and major river systems and those that were not. Empirical arguments can be made to explain why submarine landslides around the world appear to be more common at the end of the last glacial maximum (LGM) and the beginning of the Holocene (~20,000 to 7,000 years ago) than at present. These include the amount of sediment reaching the slope, expected to be higher during low sea level when rivers and glaciers discharged at the shelf edge; the amount of sediment delivered to the coast by catastrophic draining of glacial lakes and by increased erosion due to wetter conditions at the LGM; and the hypothesized increase of seismicity resulting from the unloading of the crust by the melting ice caps (Lee, 2009) and from the sediment and water loading of the margin (Brothers et al., submitted). Pore pressure increase in slope sediments may have accompanied the rapid accumulation of sediments on the continental slope and rise (Flemings et al., 2008; Dugan and Flemings, 2000; and Kvalstad et al., 2005).

Relative dating of landslides can be determined by the cross-cutting relationships between landslides scars and submarine canyons (Chaytor et al., 2012). In the Atlantic margin we observed scars that have been dissected by canyons while others have blocked canyon flows. Our recent work indicates that most submarine canyons along the Atlantic margin are relict and last active at the end of the last sea level lowstand, but some may still be active to some degree today. A better understanding of the oceanographic and sedimentological conditions required to incise submarine canyons (e.g., Brothers et al., in revision) will help to date both canyon activity and the landslides they intersect.

Because the vast majority of landslide tsunamis are associated with earthquakes, and because tsunami height is scaled with landslide volume, earthquake probability may be used to estimate maximum landslide tsunami probability. Volume and area distributions of submarine landslides along the Atlantic margin follow lognormal-like or double-Pareto distribution (Chaytor et al., 2009, and ten Brink et al., 2012). These lognormal-like distributions can be simulated (both underwater and on land) using slope stability analysis and the expected peak spectral acceleration as a function of distance from the earthquake source (ten Brink et al., 2009). Therefore, the maximum area and volume of a landslide appears to be related to the magnitude of the triggering earthquake. This approach predicts that earthquake magnitudes < 4.5 and earthquakes located more than 100 to 150 km from the continental slope are generally incapable of generating landslides. This approach can be applied to passive margins with clastic (sand, clay) sediments. Some U.S. margins, such as around Puerto Rico and Florida north to South Carolina, are composed predominantly of carbonate material. These margins are characterized by steep slopes (≤45°), reflecting the strong cohesion of carbonate rocks. Landslide distribution along the carbonate rock margin of Puerto Rico was found to be exponential, not lognormal-like (ten Brink et al., 2006). This distribution can be explained if carbonate rocks are weakened by fissures that have formed by dissolution. Fissure distribution is random. Therefore landslide size will be determined by the available block size during earthquake shaking, which is random, not by the earthquake magnitude. While many seismically active margins are also covered by clastic sediments, landslides in these margins may often be much smaller than those predicted from the earthquake magnitude because the frequency of shaking outpaces the rate of sediment accumulation along the margin. However, to date we are not aware of a regional landslide distribution study for a seismically active margin.

Earthquakes are the result of slow stress accumulation, of faults that are preferentially oriented within the stress field, and sometimes of rheological and compositional conditions within the fault zone. Analogous conditions exist for landslides. First, for a landslide to occur, unconsolidated sediment should be available to fail. It is therefore likely that landslide distribution is not uniform in space but is concentrated in regions with thick unconsolidated sediments. Along the Atlantic margin it appears that most landslides have occurred offshore of New York and New England, where glacial processes supplied sediment to the margin, and offshore of the Mid-Atlantic region where LGM delta fronts are located (Twichell et al., 2009). Because the location and size of landslides depends on sediment availability, one can argue that areas that have already failed will not fail again as long as additional sediments are not introduced to the margin. Such an argument could help identify high-hazard vs. low-hazard sections of the margin. Landslide history in Norway indicates that large landslides such as the Storegga landslide ~7500 years ago have also occurred at the end of the previous glacial maximum, but not in the intervening period or since ~7500 years ago (Halfidason et al., 2005). However, to date there is no systematic study that could confirm this potentially important prediction. A comparison of size distribution of landslides in passive and active margins, where seismic activity outpaces sediment supply, may provide some constraints.

Slope stability decreases with increasing slope gradient. Some margins are steeper than others owing to pre-existing physiography over which the sediments were deposited. For example, the New England-Georges Banks continental slope is steeper than slopes farther south along the margin because of Mesozoic reefs forming "a wall" at shallow depths beneath the slope (Brothers et al., 2013). This must have contributed to the prevalence of landslides in that sector of the margin (Twichell et al., 2009). Stability also decreases with increasing pore pressure. Therefore, landslides can occur at slopes as low as 0.1° to 0.5° (off the mouth of the Mississippi Delta, Prior and Coleman, 1978). Large landslide scars are also observed on the open slope of the U.S. Atlantic continental rise where the sea floor has gradients of 1° to 2° (Twichell et al., 2009). However, to date there has not been a systematic mapping of pore pressure with sediments in the margin to determine the spatial dimensions of high pore-pressure regions. It is unclear whether these dimensions are larger than a kilometer (Sultan et al., 2010). It is possible that canyon incision of the continental slope may lower the regional pore pressure there. Gas venting has been recently detected along canyon walls (http://www.noaanews.noaa.gov/stories2012/20121219_gas_seeps.html), but it is presently unclear how widespread this venting is.

Phrampus and Hornbach (2012) have proposed that changes in the Gulf Stream in the past 5,000 years had caused widespread gas-hydrate destabilization, which perhaps caused the Cape Fear slide. Based on clustering of many head scarps along the Atlantic margin at the expected upper water depth of the gas hydrate stability zone, Booth et al. (1993) proposed that gas hydrate dissociation promotes slope failure. However, the age of the Cape Fear slide had been estimated to be between 27,000 and 10,000 years ago (Rodriguez and Paull, 2000) and our newer high-resolution bathymetry maps show no clustering of the head walls at the 800-m depth (Twichell et al., 2009, and ten Brink et al., 2012). The depth range of the majority of the mapped head scarps is 1000 to 1900 m. Thus, gas-hydrate dissociation cannot be linked directly to the generation of landslides, although it may contribute to increased pore pressure in some locations.

The rise of salt diapirs is expected to increase the slope gradient in certain locations and may even cause slope instability. Rising salt diapirs in the area of Cape Fear and Cape Lookout could have destabilized the slope there, but causal connection has so far not been established. Some landslides in the Gulf of Mexico, such as the large East Break and DeSoto slides are located in areas of salt diapirs. Earthquakes might not have been the cause of these landslides because the seismicity rate is very low and historical earthquake magnitudes are small (< M6).

An additional challenge to estimating landslide tsunami probability results from the uncertainty in the coupling of energy between the sliding mass and the water column. Tsunamis from two landslides with identical volumes could have different amplitudes depending on the slide speed and bottom friction. Moreover, the suggestion that, in passive margins with clastic sediments, landslide area is related to earthquake magnitude implies that a landslide is in fact an aggregate of many small slope failures within an area that was subject to a supercritical horizontal acceleration. The failure area of the 1929 Grand Banks landslide, which caused a devastating tsunami, exhibits patches of failures interspersed with seafloor patches where no failure was detected (Mosher and Piper, 2007). Both the observation of the 1929 slide scar and the suggested linkage between landslide size and earthquake magnitude call into question the mechanism by which a tsunami is excited from an aggregate failure: Is the tsunami the result of constructive interference of many small failures? Or is it generated when convergence of debris flows converge in existing canyon and valley corridors and become several hundreds of meters thick? Or is it generated when a debris flow ignites into turbidity flow as it undergoes a hydraulic jump? More research is needed into this fundamental question.

References

Booth, J.S., O'Leary, D.W., Popenoe, P., Danforth, W.W., 1993, "U.S. Atlantic continental slope landslides: Their distribution, general attributes, and implications," *in*, Schwab, W.C., Lee, H.J., and Twichell, D.C., eds., "Submarine landslides: Selected studies in the U.S. Exclusive Economic Zone," U.S. Geological Survey Bulletin No. 2002, pp. 14–22.

Brothers, D.S., ten Brink, U.S., Andrews, B.D. & Chaytor, J.D., April 2013, "Geomorphic characterization of the U.S. Atlantic continental margin," *Mar. Geol.* Vol. 338; pp. 46–63.

Brothers, D.S., ten Brink, U.S., Andrews, B.D., Chaytor, J.D., & Twichell, D.C., February 2013, "Sedimentary process flow fingerprints in submarine canyons," *Mar. Geol.* Vol. 337; pp. 53–66.

Brothers, D.S., Luttrell, K.M., Chaytor, J.D., July 2013, "Sea level induced seismicity and submarine landslide occurrence," *Geology*, Geological Society of America.

Chaytor, J.D., Twichell, D.C., and ten Brink, U.S., 2012, "A Reevaluation of the Munson-Nygren-Retriever Submarine Landslide Complex, Georges Bank Lower Slope, Western North Atlantic," In, Yamada, Y. (ed.), *Submarine Mass Movements and Their Consequences, Advances in Natural and Technological Hazards Research* v. 31, Springer, New York, pp. 135–145.

Chaytor, J. D., ten Brink, U. S., Solow, A. R., and Andrews, B. D., 2009, "Size distribution of submarine landslides along the U.S. Atlantic Margin and its implications to tsunami hazards," *Mar. Geol.* 264:16–27.

Corral, A., 2004.,"Long-term clustering, scaling, and universality in the temporal occurrence of earthquakes," *Phys. Rev. Lett.* 92, doi:10.1103/PhysRevLett.1192.108501.

Dugan, B., and Flemings, P.B., 2000, "Overpressure and fluid flow in the New Jersey continental slope: implications for failure and cold seeps," *Science*, vol. 289:288–291.

Flemings, P. B., Long, H., Dugan, B., Germaine, J., John, C. M., Behrmann, J. H., Sawyer, D. & IODP Expedition, 308 Scientists, 2008, "Pore pressure penetrometers document high overpressure near the seafloor where multiple submarine landslides have occurred on the continental slope, offshore Louisiana, Gulf of Mexico," *Earth Planet. Sci. Lett.* 269:309-325.

Haflidason, H., R. Lien, H. P. Sejrup, C. F. Forsberg, and P. Bryn, 2005, "The dating and morphometry of the Storegga Slide," *Mar. Petrol. Geol.* 22:123–136.

Kagan, Y.Y., 2010, "Statistical distributions of earthquake numbers: consequence of branching process," *Geophys. J. Int.* 180:1313–1328.

Kvalstad, T.J., Andresen, L., Forsberg, C.F., Berg, K., Bryn, P., and Wangen, M., 2005, "The Storegga slide: evaluation of triggering sources and slide mechanics," *Mar. Petrol. Geol.* 22:244–256.

Lee, H.J., 2009, "Timing of occurrence of large submarine landslides on the Atlantic ocean margin," *Mar. Geol.* 53–64.

Mosher, D., and Piper, D., 2007, "Analysis of multibeam seafloor imagery of the Laurentian Fan and the 1929 Grand Banks landslide area," In, Lykousis, V., et al. (eds.), *Submarine Mass Movements and Their Consequences*, v. 27, Springer, New York, pp. 77–88.

Parsons, T., 2002, "Global Omori law decay of triggered earthquakes: Large aftershocks outside the classical aftershock zone," *J. Geophys. Res.* 107:2199, doi:2110.1029/2001JB000646.

Phrampus, B. J., and Hornbach, M. J., 2012, Recent changes to the Gulf Stream causing widespread gas hydrate destabilization. Nature, 490(7421), 527-530.

Prior, D. B., and Coleman, J. M., 1978, "Disintegrating retrogressive landslides on very-low-angle subaqueous slopes, Mississippi delta," *Marine Georesources and Geotechnology* 3:37–60.

Sultan, N., Marsset, B., Ker, S., Marsset, T., Voisset, M., Vernant, A. M., Bayon, G., Cauquil, E., Adamy, J., Colliat, J.L & Drapeau, D., 2010, Hydrate dissolution as a potential mechanism for pockmark formation in the Niger delta," *J. Geophys. Res.* 115:B08101, doi: 10.1029/2010JB007453.

Swafford, L., and Stein, S., 2007, "Limitations of the short earthquake record for seismicity and seismic hazard studies," *Special Papers of the Geological Society of America* 425:49–58.

Rodriquez, N. M., & Paull, C. K., 2000, "^{14}C dating of sediments of the uppermost Cape Fear slide plain: Constraints on the timing of this massive submarine landslide," In *Proceedings ODP, Scientific Results* 164:325–327, College Station, TX (Ocean Drilling Program).

ten Brink, U.S., J.D. Chaytor; B.D. Andrews, D.S. Brothers, E.L. Geist, 2012, "Updated size distribution of submarine landslides along the U.S. Atlantic margin," *AGU* 90(52), Fall Meet. Suppl. Abstract OS43C-1827.

ten Brink, U.S., Barkan, R., Andrews, B.D., Chaytor, J.D., 2009, "Size distributions and failure initiation of submarine and subaerial landslides," *Earth Planet. Sci. Lett.* 287:31–42.

ten Brink, U.S., Geist, E.L., Andrews, B.D., 2006, "Size distribution of submarine landslides and its implication to tsunami hazard in Puerto Rico," *Geophys. Res. Lett.* 33, doi:10.1029/2006GL026125.

Twichell, D. C., Chaytor, J. D., ten Brink, U. S., and Buczkowski, B., 2009, "Geologic Controls on the Distribution of Submarine Landslides along the U.S. Atlantic Continental Margin," *Mar. Geol.*, 4-15.

6.7.4 Tsunami Flooding Assessment Using Forecast Tools

Yong Wei[1, 2], Vasily Titov[1]

[1]Pacific Marine Environmental Laboratory, NOAA, Seattle, WA 98115
[2]Joint Institute for the Study of Atmosphere and Ocean, University of Washington, Seattle, WA 98115

NOAA's Pacific Marine Environmental Laboratory is developing tsunami modeling tools as part of the realtime tsunami forecast system for NOAA's Tsunami Warning Centers. The models are used in combination with the realtime deep-ocean measurements to produce estimates of tsunami parameters for coastal locations before the wave reaches the coast (Titov, 2005 and 2009; Tang et al., 2009 and 2012; and Wei et al., 2008 and 2012). This realtime tsunami hazard assessment will help to provide an informative and site-specific warning for coastal communities. Combined with education and mitigation measures, the tsunami forecast and warning will provide an effective means for coastal communities to prevent loss of lives from tsunamis. It will also reduce the chances for unnecessary evacuations caused by over-warning.

The modeling tools that have been developed for the realtime forecast could also be used for the long-term tsunami hazard assessment, in terms of a deterministic (Tang et al., 2009, and Uslu et al., 2010) or probabilistic (González et al., 2009) approach. The forecast models for oceanwide tsunami propagation and coastal inundation are thoroughly developed and tested to provide the best possible accuracy. These models provide an opportunity for unprecedented quality scope of tsunami hazard assessment for a particular community along the U.S. Pacific and Atlantic coastlines. Together with Pacific Marine Environmental Laboratory's (PMEL's) model database of tsunami propagation, these models are able to relate the PTHA offshore wave height to onshore flooding zones for tsunami hazards associated with a certain design return period. PMEL is collaborating with URS and the American Society of Civil Engineers (ASCE) to explore methodologies to develop 2,500-year tsunami flooding zones based on a maximum tsunami amplitude of 30 m depth obtained through PTHA and their disaggregated tsunami sources.

Several examples of tsunami hazard assessments using the forecast tools will be presented.

References

González, F.I., E.L. Geist, B. Jaffe, U. Kânoğlu, H. Mofjeld, C.E. Synolakis, V.V. Titov, D. Arcas, D. Bellomo, D. Carlton, T. Horning, J. Johnson, J. Newman, T. Parsons, R. Peters, C. Peterson, G. Priest, A. Venturato, J. Weber, F. Wong, and A. Yalciner (2009): "Probabilistic tsunami hazard assessment at Seaside, Oregon, for near- and far-field seismic sources," *J. Geophys. Res.* 114:C11023, doi: 10.1029/2008JC005132.

Tang, L., V.V. Titov, E. Bernard, Y. Wei, C. Chamberlin, J.C. Newman, H. Mofjeld, D. Arcas, M. Eble, C. Moore, B. Uslu, C. Pells, M.C. Spillane, L.M. Wright, and E. Gica (2012), "Direct energy estimation of the 2011 Japan tsunami using deep-ocean pressure measurements," *J. Geophys. Res.* 117:C08008, doi: 10.1029/2011JC007635.

Tang, L., V. V. Titov, and C. D. Chamberlin (2009), "Development, testing, and applications of site-specific tsunami inundation models for real-time forecasting," *J. Geophys. Res.* 114:C12025, doi:10.1029/2009JC005476.

Titov, V.V. (2009), "Tsunami forecasting," Chapter 12 in *The Sea, Volume 15: Tsunamis*, Harvard University Press, Cambridge, MA, and London, England, pp. 371–400.

Titov, V.V., F.I. González, E.N. Bernard, M.C. Eble, H.O. Mofjeld, J.C. Newman, and A.J. Venturato (2005), "Real-time tsunami forecasting: Challenges and solutions," *Nat. Hazards* 35(1):41–58, Special Issue, U.S. National Tsunami Hazard Mitigation Program.

Uslu, B., V.V. Titov, M. Eble, and C. Chamberlin (2010), "Tsunami hazard assessment for Guam," NOAA OAR Special Report, Tsunami Hazard Assessment Special Series, Vol. 1, National Oceanic and Atmospheric Administration, Washington, DC.

Wei, Y., C. Chamberlin, V.V. Titov, L. Tang, and E.N. Bernard (2012), "Modeling of 2011 Japan Tsunami - lessons for near-field forecast," *Pure Appl. Geophys.*, doi:10.1007/s00024-012-0519-z

Wei, Y., E. Bernard, L. Tang, R. Weiss, V. Titov, C. Moore, M. Spillane, M. Hopkins, and U. Kânoğlu (2008), "Real-time experimental forecast of the Peruvian tsunami of August 2007 for U.S. coastlines," *Geophys. Res. Lett.* 35:L04609, doi:10.1029/2007GL032250.

6.7.5 Probabilistic Tsunami Hazard Mapping in California Feasibility and Applications

Charles R. Real[1] and Rick Wilson[1]

[1]California Geological Survey, Sacramento, CA 95814

As part of California's Tsunami Hazard Mitigation and Preparedness Program, the California Geological Survey (CGS) is investigating the feasibility of designating official tsunami hazard zones under authority of the Seismic Hazards Mapping Act (California Public Resources Code Sec. 2731 et seq.) and the generation of other products that would facilitate tsunami hazard mitigation and risk reduction throughout coastal communities. Several pilot projects are underway which are briefly described.

Probabilistic Seismic Hazard Analysis (PSHA) is the foundation for estimating earthquake-resistant design loads for engineered structures in California and has been considered in land-use decisions at state and local levels for more than a decade. Probabilistic Flood Hazard Analysis (PFHA) forms the basis for the National Flood Insurance Program's 100- and 500-year flood maps and California's 200-year flood hazard maps (California Government Code Sec 65300.2 et seq.). A probabilistic approach provides a more quantitative analysis for risk-based decisionmaking when considering likelihood of loss and actions that can be taken to reduce it. With a sound theoretical basis (Geist and Parsons, 2006) and recent advances in probabilistic tsunami inundation modeling (Thio, 2010, and González et al., 2009), it is timely to consider application of Probabilistic Tsunami Hazard Analysis (PTHA) to risk reduction. For example, the American Society of Civil Engineer's Subcommittee on Tsunami Loads and Effects is currently considering 100-year and 2500-year events in the development of recommended tsunami-resilient design provisions for the 2018 International Building Code (ASCE, 2012). However, recent worldwide events have brought into question the reliability of PSHA, particularly considering the long, variable, and uncertain recurrence times of large earthquakes relative to exposure times. High accelerations are often occurring in areas shown as low on hazard maps, even for locations having a long seismic history (Peresan and Panza, 2012; Stein et al., 2011). Considering difficulties in estimating tsunami frequency brings to question the viability of PTHA, and highlights the need to thoroughly assess the modeling framework and reliability of input data before derivative products can have application to public policy.

With support from the National Tsunami Hazard Mitigation Program (NTHMP), CGS and the California Emergency Management Agency are partnered with URS Corporation, the Pacific Earthquake Engineering Research Center, and the California Department of Transportation (Caltrans) to develop probabilistic tsunami hazard maps for the entire California coastline. Initial PTHA results are currently used by Caltrans (a project sponsor for URS work) to evaluate the vulnerability of coastal transportation facilities, while also serving as a means to explore prototype products for local land-use planning applications. Work completed thus far has focused on the development of a practical PTHA procedure that can be applied to a large region and refinement of nonlinear wave propagation methods for estimating onshore wave heights, current velocities, and inundation levels from distant sources (Thio and Sommerville, 2010). This work is currently expanding to include the Cascadia Subduction zone and smaller local sources offshore of southern and central California. The NTHMP Mapping and Modeling Subcommittee has recommended California's program be considered a "national pilot" for PTHA mapping, and is actively pursuing including this work in the FY2013–2017 NTHMP Strategic Plan. To better understand limitations and uncertainties in PTHA, results from the State project for Crescent City, California are being compared with those of an independent, FEMA-sponsored analysis of Crescent City by Baker/AECOM and the University of Washington. That project is evaluating the feasibility of incorporating tsunamis into coastal flood hazard mapping for the National Flood Insurance Program based on improvements to the methodology used in a comprehensive PTHA of Seaside, Oregon (González et al., 2009). A committee of experts has been assembled that will evaluate results from the two teams considering the differences in source, propagation, and inundation components and probabilistic framework of each model.

A parallel pilot project is exploring issues related to implementation of PTHA and derivative hazard mitigation products. Application of conceptual and prototype products and associated policies are being discussed with planning and public works departments in Crescent City, a small community with a weak economy, and Huntington Beach, a large affluent community. Results are expected to help identify viable mitigation products and policy adjustments that may be necessary for tsunami hazard mitigation to work given the socioeconomic diversity among California's coastal communities.

A third project supported under the FEMA's RISKMAP program is developing high-resolution prototype products for the maritime sector based on high-resolution hydrodynamic modeling of tsunami induced surge in five California harbors: San Diego Port and Harbor, Ports of Los Angeles and Long Beach, Ventura Harbor, Santa Cruz Harbor, and Crescent City Harbor. The goal is to model wave heights and current flow in order to identify zones of high current velocity that can be used to develop navigation guides for evacuation and for strengthening port infrastructure to reduce tsunami impact. It is anticipated that products will be derived from both deterministic and probabilistic hazard analyses.

In addition to the expert PTHA evaluation panel, California has established a Tsunami Policy Working Group, operating under the California Natural Resources Agency, Department of Conservation, which is composed of experts in earthquakes, tsunamis, flooding, structural and coastal engineering, and natural hazard policy from government, industry, and nonprofit natural hazard risk-reduction organizations. The working group serves a dual purpose, being an advisor to the State tsunami program and a consumer of insights from the Science Application for Risk Reduction (SAFRR) Tsunami Scenario project. The latter is a USGS study to evaluate the impact of a magnitude 9.1 mega-thrust earthquake occurring along the Aleutian Islands Subduction Zone that presents the greatest distant tsunami threat to southern and central California. The working groups' role is to identify, evaluate, and make recommendations to resolve issues that are preventing full and effective tsunami hazard mitigation and risk reduction throughout California's coastal communities. Committee membership is selected to represent entities responsible for coastal development, insurance, local and regional planning, public works, foreign and domestic disaster preparedness, recovery and seismic policy. Among those selected are representatives of the two cities chosen for the State's tsunami pilot project. Their participation in working group deliberations provides opportunity to bring forth local implementation issues in a forum conducive to multidisciplinary resolution, and an opportunity to incorporate a local perspective while formulating recommendations.

Finally, CGS is participating as an Associate member in the ASCE Subcommittee on Tsunami Loads and Effects where insights from the aforementioned projects are being brought to bear on the development of prototype products supporting the ASCE 7-2015 recommended provisions for the International Building Code. The State participated in a similar role years ago when PSHA was introduced into seismic provisions of the building code. Products under consideration are a 2,500-year tsunami inundation zone that would trigger the code process for high-risk category buildings, and production of a database of offshore wave heights and principal tsunami sources along the coast from de-aggregated PTHA that can provide the necessary input for site-specific deterministic inundation analyses used to estimate design loads for proposed construction projects. The work currently underway to evaluate and verify PTHA models is complementary to product and application development, and will facilitate California's adoption of tsunami code provisions when they become available.

References

American Society of Civil Engineers (ASCE), 2012, *Subcommittee on Tsunami Loads and Effects Workshop, July 27–28*, Portland, OR.

Geist, E.L., and Parsons, T., 2006, "Probabilistic Analysis of Tsunami Hazards," *Natural Hazards* 37:277–314.

González, F.I., Geist, E. L., Jaffe, B., Kânoğlu, U., Mofjeld, H., Synolakis, C.E., Titov, V.V., Arcas, D., Bellomo, D., Carlton, D., Horning, T., Johnson, J., Newman, J., Parsons, T., Peters, R., Peterson, C., Priest, G., Venturato, A., Weber, J., Wong, F., and Yalciner A. 2009, "Probabilistic tsunami hazard assessment at Seaside, Oregon, for near- and far-field seismic sources," *Journ. Geoph. Res.* 114:C11023.

Parasen, A., and Panza, G.F., 2012, "Improving Earthquake Hazard Assessments in Italy: an Alternative to 'Texas Sharpshooting,'" *Trans. Am. Geoph. Union* 93:538.

Stein, S., Geller, R. and Liu, M., 2011, "Bad Assumptions or Bad Luck: Why Earthquake Hazard Maps Need Objective Testing," *Seism. Res. Letters* 82:623–626.

Thio, H.K., and Sommerville, P., 2010, "Probabilistic Tsunami Hazard in California," Pacific Earthquake Engineering Research Center, PEER 2010/108.

6.7.6 Statistical Testing of Hypotheses and Assumptions Inherent to Probabilistic Tsunami Hazard Analysis (PTHA)

Eric L. Geist[1], Tom Parsons[1], and Uri ten Brink[2]

[1]U.S. Geological Survey, Menlo Park, CA 94025
[2]U.S. Geological Survey, Woods Hole, MA 02543

Probabilistic Tsunami Hazard Analysis (PTHA) methods have recently been derived from the well-established Probabilistic Seismic Hazard Analysis (PSHA) method that calculates the probability of ground shaking from earthquakes, as originally developed by Cornell (1968). Both PSHA and PTHA involve three basic steps: (1) source characterization, including definition of source parameter distributions and recurrence rates; (2) calculation of wave propagation and attenuation effects from source to site; and (3) aggregation of hazard probabilities at the site from all sources considered in the analysis (e.g., Geist and Parsons, 2006; Geist et al., 2009). The primary differences between PTHA and PSHA are that distant sources must be considered in PTHA, owing to the slow attenuation of tsunami waves in the ocean, and that numerical propagation models can be used in PTHA, in place of empirical attenuation relationships used in PSHA. The standard forms of PTHA involve assumptions, such as sources occurring as a Poisson process, which are often untested in practice. Although different hypotheses representing epistemic uncertainty can be incorporated into PTHA using a logic-tree framework, there are a number of statistical tools available with which we can possibly exclude or confirm certain hypotheses, thus reducing epistemic uncertainty. We focus here on statistical tests of recurrence distributions, size distributions, dependence among parameters, and the ergodic hypothesis. Where standard assumptions may be found to be invalid, new approaches to PTHA need to be developed.

It is often assumed that the recurrence distribution of sources follows an exponential distribution associated with a stationary Poisson process. For earthquakes, the inclusion of aftershocks and triggered events results in temporal clustering at a greater degree than represented by an exponential distribution (e.g., Parsons, 2002; Corral, 2004; and Kagan, 2010). For very large earthquakes (M≥8.3), non-Poissonian behavior has not been observed, given the amount and uncertainty in available earthquake data (Parsons and Geist, 2012). Temporal clustering of tsunamis themselves has been demonstrated and tested against a Poisson null hypothesis (Geist and Parsons, 2011), though aftershocks *sensu stricto* account for only a part of the overabundance of short inter-event times. Non-Poissonian recurrence distributions can approximately be incorporated into standard PTHA methodology, using an apparent Poisson rate parameter (Petersen et al., 2007). For landslide sources, there is very little data to test the Poisson hypothesis, although if the number of landslide events can be determined from marine geophysical data and a basal horizon can be dated, Bayesian techniques can be used to determine the most likely rate parameter and its uncertainty (Geist and Parsons, 2010; Geist and ten Brink, 2012). Optimally, geologic age-dates of landslides from drill hole samples or cores can be used to more accurately obtain recurrence rates and assess various probability models using, for example, Akaike's information criterion (Geist et al., in review). Unfortunately, there are very few locations where such data are available. Another issue with landslide tsunamis is the long-term dependence of submarine landslide activity on glacial cycle and sea-level rise (Lee, 2009). Nonstationary source rates have yet to be incorporated into PTHA.

There has been considerable discussion regarding distribution of tsunami source sizes, particularly for earthquakes. Competing hypotheses include the characteristic earthquake model, in which the largest earthquakes are defined by fault segmentation and the historical record, and the Gutenberg-Richter earthquake model, in which earthquake sizes follow a Pareto distribution (cf. Parsons and Geist, 2009). Although the characteristic earthquake model is commonly used in both PSHA and PTHA, a number of studies refute this model (e.g., Kagan and Jackson, 1995; Rong et al., 2003; Parsons et al., 2012), particularly for subduction-zone earthquakes that generate the majority of the world's tsunamis. Submarine landslides also tend to follow a Pareto distribution like their on-land counterparts (ten Brink et al., 2006), although in certain environments, it has been shown that a log-normal distribution is a more appropriate model (ten Brink et al., 2009). The Pareto distribution for source sizes can, in general, be considered the null hypothesis; other distributions can be tested against the Pareto distribution using techniques, for example, reviewed by Clauset et al. (2009).

Independence is often assumed in PTHA and PSHA among source parameters and their uncertainty. It is often difficult to develop methods to accommodate dependent parameters in probabilistic analysis. An example in PTHA is combining near-shore tsunami waveforms with tidal variations: a source of aleatory uncertainty, because tsunami sources cannot be predicted in time. The method of Mofjeld et al. (2007) used in the FEMA Seaside PTHA Pilot Study (Tsunami Pilot Study Working Group, 2006), uses the linear combination of the tsunami amplitude envelope with site-specific distribution of tidal heights. However, landslide tsunami waves can be significantly nonlinear, such that differences in antecedent sea level cannot simply be added or subtracted to tsunami wave heights computed at a given vertical datum. Copula methods, common in many hydrology applications, can be applied to examine this and other dependent structures in PTHA. Page and Carson (2006) present an application of copula methods in determining earthquake probabilities given uncertainties in the data and models.

Finally, it can be very useful to assume that spatial variations in source characteristics are equivalent to temporal variations at a point under the ergodic hypothesis (Geist et al., 2009). For example, in the absence of drill-hole data that samples submarine landslides throughout geologic time, one can date submarine landslides expressed at the sea floor over a specified region, yielding an occurrence rate for the entire region. However, this assumes a similar geologic framework (e.g., clastic vs. carbonate) and morphology (e.g., canyon vs. slope) throughout the region. Statistical methods to test whether or not the ergodic hypothesis holds under specific conditions (cf. Anderson and Brune, 1999) need to be developed in the context of PTHA.

References

Anderson, J.G., Brune, J.N., 1999, "Probabilistic seismic hazard analysis without the ergodic assumption," *Seismol. Res. Lett.* 70:19–28.

Clauset, A., Shalizi, C.R., Newman, M.E.J., 2009, "Power-law distributions in empirical data," *SIAM Review* 51:661–703.

Cornell, C.A., 1968, "Engineering seismic risk analysis," *Bull. Seismol. Soc. Am.* 58:1583-1606.

Corral, A., 2004, "Long-term clustering, scaling, and universality in the temporal occurrence of earthquakes," *Physical Review Letters* 92, doi: 10.1103/PhysRevLett.1192.108501.

Geist, E.L., Chaytor, J.D., Parsons, T., ten Brink, U., April 2013. "Estimation of submarine mass failure probability from a sequence of deposits with age dates," *Geosphere*, Vol. 9; pp. 287-298.

Geist, E.L., Parsons, T., 2006, "Probabilistic analysis of tsunami hazards," *Natural Hazards* 37:277–314.

Geist, E.L., Parsons, T., 2010, "Estimating the empirical probability of submarine landslide occurrence," in: Mosher, D.C., Shipp, C., Moscardelli, L., Chaytor, J., Baxter, C., Lee, H.J., Urgeles, R. (Eds.), *Submarine Mass Movements and Their Consequences IV*, Springer, Heidelberg, Germany, pp. 377–386.

Geist, E.L., Parsons, T., 2011, "Assessing historical rate changes in global tsunami occurrence," *Geophys. J. Int.* 187:497–509.

Geist, E.L., Parsons, T., ten Brink, U.S., Lee, H.J., 2009, "Tsunami Probability," in: Bernard, E.N., Robinson, A.R. (eds.), *The Sea*, v. 15. Harvard University Press, Cambridge, MA, pp. 93–135.

Geist, E.L., ten Brink, U.S., 2012, "NRC/USGS Workshop Report: Landslide Tsunami Probability," p. 635.

Kagan, Y.Y., 2010, "Statistical distributions of earthquake numbers: consequence of branching process," *Geophys. J. Int.* 180:1313–1328.

Kagan, Y.Y., Jackson, D.D., 1995, "New seismic gap hypothesis: Five years after," *J. Geophys. Res.* 100:3943–3959.

Lee, H.J., 2009, "Timing of occurrence of large submarine landslides on the Atlantic ocean margin," *Mar. Geol.*, 53–64.

Mofjeld, H.O., González, F.I., Titov, V.V., Venturato, A.J., Newman, A.V., 2007, "Effects of tides on maximum tsunami wave heights: Probability distributions," *Journal of Atmospheric and Oceanic Technology* 24:117–123.

Page, M.T., Carlson, J.M., 2006, "Methodologies for earthquake hazard assessment: Model uncertainty and the WGCEP-2002 forecast," *Bull. Seismol. Soc. Am.* 96:1624–1633.

Parsons, T., 2002, "Global Omori law decay of triggered earthquakes: Large aftershocks outside the classical aftershock zone," *J. Geophys. Res.* 107:2199, doi:2110.1029/2001JB000646.

Parsons, T., Console, R., Falcone, G., Murru, M., Yamashina, K., 2012, "Comparison of characteristic and Gutenberg-Richter models for time-dependent M ≥ 7.9 earthquake probability in the Nankai-Tokai subduction zone, Japan," *Geophys. J. Int.*, doi: 10.1111/j.1365-1246X.2012.05595.x.

Parsons, T., Geist, E.L., 2009, "Is there a basis for preferring characteristic earthquakes over a Gutenberg-Richter distribution in probabilistic earthquake forecasting?", *Bull. Seismol. Soc. Am.* 99:2012–2019.

Parsons, T., Geist, E.L., 2012, "Were global M≥8.3 earthquake time intervals random between 1900-2011?", *Bull. Seismol. Soc. Am.* 102, doi:10.1785/0120110282.

Petersen, M.D., Cao, T., Campbell, K.W., Frankel, A.D., 2007, "Time-independent and time-dependent seismic hazard assessment for the State of California: Uniform California Earthquake Rupture Forecast Model 1.0.," *Seismol. Res. Lett.* 78:99–109.

Rong, Y., Jackson, D.D., Kagan, Y.Y., 2003, "Seismic gaps and earthquakes," *J. Geophys. Res.* 108:ESE 6-1 through 6-14.
ten Brink, U.S., Barkan, R., Andrews, B.D., Chaytor, J.D., 2009, "Size distributions and failure initiation of submarine and subaerial landslides," *Earth Planet. Sci. Lett.* 287:31–42.

ten Brink, U.S., Geist, E.L., Andrews, B.D., 2006, "Size distribution of submarine landslides and its implication to tsunami hazard in Puerto Rico," *Geophys. Res. Lett.* 33, doi:10.1029/2006GL026125.

Tsunami Pilot Study Working Group, 2006, "Seaside, Oregon Tsunami Pilot Study - Modernization of FEMA Flood Hazard Maps," USGS Open-File Report 2006-1234, U.S. Geological Survey, Reston, VA.

PANEL 6

RIVERINE FLOODING

Co-Chairs:
Will Thomas, Michael Baker, Jr., Inc.
Rajiv Prasad, PNNL

Technical Reporters:
Peter Chaput and Jeff Mitman, NRC

7.   Riverine Flooding

Will Thomas[1], Rajiv Prasad[2], Peter Chaput[3] and Jeff Mitman[4]

[1]Michael Baker, Jr., Inc., Manassas, VA
[2]Pacific Northwest National Laboratory, Richland, WA
[3]Office of New Reactors, U.S. NRC, Rockville, MD
[4]Office of Nuclear Reactor Regulation, U.S. NRC, Rockville, MD

7.1   Motivation

Panel 6 on Riverine Flooding focused on riverine flooding, including watershed responses through routing precipitation events and accounting for antecedent conditions such as snowpack releases. The Probabilistic Flood Hazard Assessment (PFHA) Workshop focused on procedures for estimating extreme flood hazards with exceedance probabilities up to 10^{-6}.

7.2   Background

The Riverine Flooding session focused on how watershed modeling processes and other ways of extending observed data can be used to estimate the exceedance probabilities of extreme flood discharges. The watershed modeling approach uses a wide range and combination of input data such as precipitation, antecedent soil moisture, snowpack depth and density and initial reservoir levels to estimate thousands of years of flood data. Paleoflood techniques use data dating from thousands of years prior to observed data to estimate the exceedance probabilities of extreme flood discharges. For regulated watersheds, hydrologic routing techniques and ratios of observed discharges can be used to estimate the exceedance probabilities of extreme flood discharges. These techniques were discussed in this session.

7.3   Overview of Presentations

There were five presentations in Panel 6:

*   Riverine PFHA for NRC Safety Reviews – Why and How? - Rajiv Prasad, PNNL

*   Flood Frequency of a Regulated River: The Missouri River – Douglas Clemetson, USACE

*   Extreme Floods and Rainfall-Runoff Modeling with the Stochastic Event Model (SEFM) – Mel Schaefer, MGS Engineering

*   Use of Stochastic Event Flood Model and Paleoflood Information to Develop Probabilistic Flood Hazard Assessment for Altus Dam, Oklahoma – Nicole Novembre, BoR

*   Paleoflood Studies and their Application to Reclamation Dam Safety – Ralph Klinger, BoR (presentation given by John England, BoR)

Rajiv Prasad defined the current status of PFHA for Nuclear Regulatory Commission projects and described the current guidance for assessing flood hazards at these projects. Doug

Clemetson described the development of unregulated and regulated frequency curves on the Missouri River and how ratios of observed floods were used to extend the regulated frequency curves. Mel Schaefer described how Monte Carlo simulation can be used to simulate thousands of years of flood hydrographs using the Storm Event Flood Model (SEFM). Nicole Novembre described the use of the SEFM and paleoflood data for estimating flood discharges up to exceedance probability of 10^{-5} on the North Fork Red River near Altus Dam in Oklahoma. John England, substituting for Ralph Klinger, described how paleoflood data were used to evaluate flood hazards at BoR dams.

7.4   Summary of Panel Discussion

As described in the presentations by Mel Schaefer and Nicole Novembre, a current approach to estimating extreme (10^{-5} or 10^{-6}) flood discharges is through the use of SEFM. Although emphasis in this workshop is on extreme floods, a better understanding is needed of less extreme floods in the range of 10^{-3} to 10^{-4}. This process involves conducting several thousand flood simulations using hydrometeorological inputs and watershed model parameters obtained from historical records within the climatic region. These hydrometeorological inputs/parameters include basin-average precipitation-frequency relationship, seasonality of the storms, spatial and temporal patterns for precipitation, antecedent soil moisture, snowpack depth and density, air temperature and freezing level and initial reservoir levels and may involve transposing hydrometeorological data from outside the watershed.

The use of SEFM and the generation of thousands of years of record from a relative small sample of observed data prompted the following questions and discussion about the accuracy of the simulated flood discharges:

- What is the information content of the simulated record? Is the resampling of a finite sample of observed events adequate in estimating flood discharges on the order of 10^{-5} and 10^{-6}? This could be delusional precision. We need to gain a better understanding of the precision of the 100,000 years of simulated record because we do not really have the equivalence of 100,000 years of record.

- The response to the above questions was that there is uncertainty in the estimated flood discharges but that the analyst needs to evaluate the observed record to determine if these data are adequate and representative of what could happen on the watershed. In some cases synthetic data should be generated or data transposed from nearby watersheds. More experience is needed with the Monte Carlo simulations in the SEFM framework. At the present time, analysts are trying to make conservative decisions based on limited data.

- Another recommendation was to estimate probability distributions for the input data like the rainfall, seasonality of the storms, antecedent moisture, water equivalent from the snowpack, and initial reservoir levels and then randomly sample from these distributions. Evaluating the joint probability of the input data may be a better approach for defining the uncertainty in the exceedance probabilities of extreme floods. This approach is similar to the coastal storm surge analyses in which probability distributions are defined for central pressure, radius to maximum winds, forward speed, angle of approach, etc. and used to define a distribution of storm surges and associated exceedance probabilities.

- The SSHAC process was mentioned and the need to have distributions of all inputs that could possibly occur. Subjective estimation of probability distributions for the input data may be needed to extend beyond observed data to obtain more scientifically valid results.

Some questions related to the hydrologic modeling process and the need for improved models. Example questions and discussion follows:

- Is the time dependence of rainfall and antecedent moisture captured adequately in the SEFM modeling approach? There may be a need to model multi-day events or to use a continuous simulation model rather than an event-type model.

- What are the limits of transferring data from nearby watershed? Physiographic similarity of the nearby watersheds needs to be considered. Are watershed models like SEFM applicable to the large watersheds? How can we capture the spatial distribution of rainfall for the very large watersheds?

- Response to the above questions indicated that SEFM has been used on watersheds up to about 8,000 square miles but the model would not be applicable for watersheds like the Missouri River on the order of 500,000 square miles. The modeler must evaluate the size of the watershed and make decisions about the spatial size and temporal distribution of storm events. This evaluation is based on historical events.

- A recommendation was to move away from the unit hydrograph method and use more physically based procedures for routing rainfall excess like the kinematic wave method.

Paleoflood data was discussed as a way of extending the observed record and adding information on floods that occurred outside the observed record. Questions and discussion related to paleoflood data follow:

- Are paleofloods that occurred several thousand years ago indicative of future flood hazards in a possibly changing climate?

- The response was that paleoflood data back to about the last ice age (~12,000 years ago) are considered reasonably consistent with current conditions given the great variability of flood data. In addition, projections of climate change have yet to reach conditions present during the Holocene Climate Optimum (5,000 to 9,000 years ago) with the warming climate following the last glacial period.

- Paleoflood analysis is currently being performed by a relatively small cadre of geomorphologists and hydrologists. There is a need for more training so that more members of the engineering community are comfortable applying these procedures.

A few comments were made about the need for coordinated operation of water-control structures in the larger watersheds. The hydrologic modeling is generally based on established operating rules but that only works if a single agency like USACE or TVA is operating the multiple reservoirs. Someone commented that FERC dams in a single large watershed are often operated by several different owners and they do not coordinate their operations. This could lead to large floods occurring that are not expected.

7.5   Observations and Insights

A major part of the Panel 6 discussion related to the accuracy of simulated floods based on a finite sample of observed events (as is used in the SEFM modeling). The opinion was that the SEFM was a good approach for estimating the exceedance probabilities of extreme (10^{-5} or 10^{-6}) floods. Possible improvements in the SEFM modeling approach included discussion about how to better account for the time dependence of rainfall and antecedent moisture conditions. The use of continuous simulation models should be considered as a way to better account for the time dependence of rainfall and antecedent moisture conditions. Based on the panel discussion there is a need for more research into the information content of 100,000 years of simulated flood discharges based on modeling a finite number of observed storm events. The general opinion was that paleoflood analysis was a good way of extending the observed record and including information on major floods outside the systematic record. Federal agencies should make the application of paleoflood methods more common practice or a part of agency policy so that the engineering community will become more comfortable with the process.

7.6   Abstracts

The workshop organizing committee developed the flooding topics to be discussed and chose the panel co-chairs for each panel topic. The co-chairs then identified potential speakers and discussed them with the organizing committee. Following agreement, the co-chairs sent out invitations to the presenters requesting presentation titles and abstracts for documentation in the workshop program. The speakers and panelists are identified in the workshop agenda (please see Appendix A). The following four abstracts document these presentations and, in some cases, reflect the discussions during the panel session.

7.6.1 Flood Frequency of a Regulated River – The Missouri River

Douglas J. Clemetson, P.E.[1]

[1]USACE Omaha District, Omaha, NE

Following the devastating floods that occurred in 1993 along the lower Missouri River and Upper Mississippi River, the Upper Mississippi River Flow Frequency Study (UMRFFS) was initiated. As part of this study, the discharge frequency relationships were updated for the Missouri River downstream from Gavins Point Dam. Establishing the discharge frequency relationships first involved extensive effort in developing unregulated flows and regulated flows for a long-term period of record at each of the main stem gaging stations. Once the unregulated and regulated hydrographs were developed, the annual peak discharges were selected for use in the discharge frequency analysis.

In order to provide a homogenous data set from which frequency analysis could be performed, effects of reservoir regulation and stream depletions had to be removed from the historic flow record. This produced a data set referred to as the "unregulated flow" data set for the period 1898–1997. A homogeneous "regulated flow" data set was also developed by extrapolating reservoir holdouts and stream depletions to present levels over the period of record 1898–1997. Flow frequency analyses were performed on the unregulated flow annual peaks using procedures found in Bulletin 17B in order to develop the unregulated flow frequency curves for the spring and summer seasons at each gage location. The spring and summer unregulated flow frequency curves were combined using the "probability of a union" equation by adding the probabilities of the spring flow frequency curve and summer flow frequency curve and subtracting the joint probability of flooding occurring from seasons to obtain the annual unregulated frequency relationships. Next, the period of record regulated flows were developed using existing reservoir regulation criteria and present level basin depletions.

The annual peaks from the regulated and unregulated data sets were then paired against each other in descending order to establish a relationship between regulated and unregulated flow at each gage location. The regulated versus unregulated relationships were extrapolated based on developing design floods by routing historic flows that were increased by as much as 100 percent. This relationship was then applied to the unregulated flow frequency curve to establish the regulated flow frequency curve. Following the record flooding in 2011, the unregulated flow and regulated flow data sets were updated to include the additional period of record 1998–2011 and comparisons were made to the past studies. It was found that the methodology utilized in the 2003 study was robust and no changes to the flow frequency relationships were required after the record flooding in 2011.

References

U.S. Army Corps of Engineers, "Hydrologic Frequency Analysis," Engineer Manual 1110-2-1415, Washington, DC, March 1993, http://publications.usace.army.mil/publications/eng-manuals/EM_1110-2-1415_sec/EM_1110-2-1415.pdf.

U.S. Army Corps of Engineers, "Upper Mississippi River Flow Frequency Study; Hydrology and Hydraulics Appendix F: Missouri River," Omaha, NB, November 2003, http://www.mvr.usace.army.mil/Portals/48/docs/FRM/UpperMissFlowFreq/App.%20F%20Omaha%20Dist.%20Hydrology_Hydraulics%20Report.pdf.

7.6.2 Extreme Floods and Rainfall-Runoff Modeling with the Stochastic Event Flood Model (SEFM)

M.G. Schaefer

MGS Engineering Consultants, Inc.

Stochastic rainfall-runoff modeling provides a methodology for development of magnitude-frequency relationships for flood peak discharge, flood runoff volume, maximum reservoir level, spillway discharge, and overtopping depths for the dam crest and spillways. The basic concept is to use Monte Carlo methods to conduct multi-thousand flood simulations utilizing hydrometeorological inputs and watershed model parameters obtained from the historical record within the climatic region. Storm magnitude is driven by a precipitation-frequency relationship for the watershed and spatial and temporal storm characteristics are simulated using storm templates developed from historical storms on the watershed or transposed from sites within the climatic region. Hydrometeorological inputs such as: storm seasonality; antecedent soil moisture; snowpack depth and density; time-series of 1000-mb air temperature and freezing level; and initial reservoir level are treated as variables rather than fixed values in the flood simulations.

The Stochastic Event Flood Model (SEFM) utilizes these concepts to develop flood-frequency relationships for flood characteristics such as flood peak discharge and maximum reservoir level for use in risk analyses for dams and other facilities. Uncertainties can also be incorporated into the flood simulations to allow development of uncertainty bounds. Development of SEFM began in 1998 and has continued over the past 15-years. During that time, SEFM has been used at over 20 dam and reservoir projects in the western United States and British Columbia, primarily to assess the likelihood of hydrologic loadings from extreme floods. This presentation will provide a basic description of the operation of SEFM and results from selected case studies.

Reference

Stocastic Event Flood Model for Hydrologic Risk Analysis:
http://www.mgsengr.com/Dam_Safety.html

7.6.3 Use of Stochastic Event Flood Model and Paleoflood Information to Develop Probabilistic Flood Hazard Assessment for Altus Dam, Oklahoma

Nicole J. Novembre, Victoria L. Sankovich, Jason Caldwell, and Jeffrey P. Niehaus

U.S. Department of Interior, Bureau of Reclamation Technical Service Center, Denver, CO

A Hydrologic Hazard Analysis (HHA) was completed at the request of the Bureau of Reclamation Dam Safety Office (Novembre et al., 2012) as one part of a Corrective Action Study for Altus Dam near Altus, OK.   This study is designed to provide flood loading and reservoir elevation information that is needed in support of a risk assessment to reduce the probability of failure under static and hydrologic loading at the dikes.

The main objectives of this study were to:

1. Collect at-site detailed paleoflood data with stratigraphy, radiocarbon ages, and hydraulic modeling within the basin, to supersede the preliminary estimates used in a 2001 study.

2. Develop a precipitation frequency relationship for the drainage area. Develop storm temporal and spatial patterns for use in a hydrologic model.

3. Develop a peak flood frequency curve using Expected Moments Algorithm (EMA).

4. Develop flood frequency relationships and inflow hydrographs with a stochastic event flood model (SEFM) (Schaefer and Barker, 2002, and MGS Engineering, 2009).

5. Route the thousands of hydrographs generated with SEFM through Altus Dam with appropriate initial reservoir water surface elevations and provide reservoir water surface elevation frequency curves and quantiles for use in risk analysis.

Paleoflood data were collected on the North Fork Red River near Altus Dam (Godaire and Bauer, 2012).   Two study reaches, one located upstream of Lake Altus and one located downstream of Altus Dam, were used to develop paleoflood data.   A non-exceedance bound was estimated with a discharge range of 120,000 ft^3/s to 141,000 ft^3/s that has not been exceeded in the last 610 to 980 years.   Three or more paleoflood events with a discharge of 26,000 ft^3/s to 36,000 ft^3/s that occurred between 790 and 1240 years ago were estimated.

A peak flood frequency curve was developed from streamflow gage data and the paleoflood data using the Expected Moments Algorithm (EMA). EMA (Cohn et al., 1997, and England et al., 2003) is a moments-based parameter estimation procedure that assumes the LP-III distribution is the true distribution for floods, and properly utilizes historical and paleoflood data.

A Stochastic Event Flood Model (SEFM) was developed for the Altus Dam watershed. Magnitude-frequency rainfall estimates, as well as rainfall storm temporal and spatial patterns, were developed as input to the model.   The model was calibrated to observed events and used to develop estimates of hydrographs for the 2% to 0.001% annual exceedance probability events. These hydrographs were routed through Altus Dam to develop reservoir water surface elevation frequency curves to be used as part of the risk analysis for Altus Dam.

The recommended hydrologic hazard curves for risk analysis are the SEFM curves using the lower and median precipitation frequency curves (Table 7–1 and Figure 7–1). The risk analysis should assume that each of these flood frequency relationships and their associated hydrographs could occur with equal probability of occurrence. The corresponding reservoir elevations (Table 7–2) were developed using reservoir operating rules provided by Reclamation's Waterways and Concrete Dams Group; therefore, the routing results can be used without modification to evaluate the potential overtopping failure mode of Altus Dam and dikes. These two curves represent equal estimates of the hydrologic hazard for baseline and risk-reduction alternatives. Extrapolation of these results should be limited to an AEP of 0.001%. The full uncertainty of the hydrologic hazard is broader than that depicted in these two curves. Uncertainty analysis should consider variations in model parameters, data reliability, and climate variability. Further consultation with a flood hydrologist is needed in the selection of inflow flood loads for corrective action.

Table 7–1 Recommended Peak Discharge and 15-day Volume Values

| Annual Exceedance Probability | Return Period (years) | Lower Precipitation Frequency | | Median Precipitation Frequency | |
		Peak Discharge (ft³/s)	15-Day Runoff Volume (ac-ft)	Peak Discharge (ft³/s)	15-Day Runoff Volume (ac-ft)
1.0%	100	40,000	150,000	48,000	181,000
0.1%	1000	85,000	329,000	108,000	419,000
0.01%	10,000	169,000	665,000	223,000	882,000
0.001%	100,000	292,000	1,098,000	390,000	1,487,000

Table 7–2 Recommended Reservoir Water Surface Elevation

| Annual Exceedance Probability | Return Period (years) | Lower Precipitation Frequency | | Median Precipitation Frequency | |
		Routing Case 1	Routing Case 2	Routing Case 1	Routing Case 2
1.0%	100	1560.7	1558.8	1561.3	1559.4
0.1%	1000	1565.2	1564.5	1567.9	1567.5
0.01%	10,000	1571.2	1571.1	1572.1	1572.1
0.001%	100,000	1572.7	1572.7	1573.4	1573.4

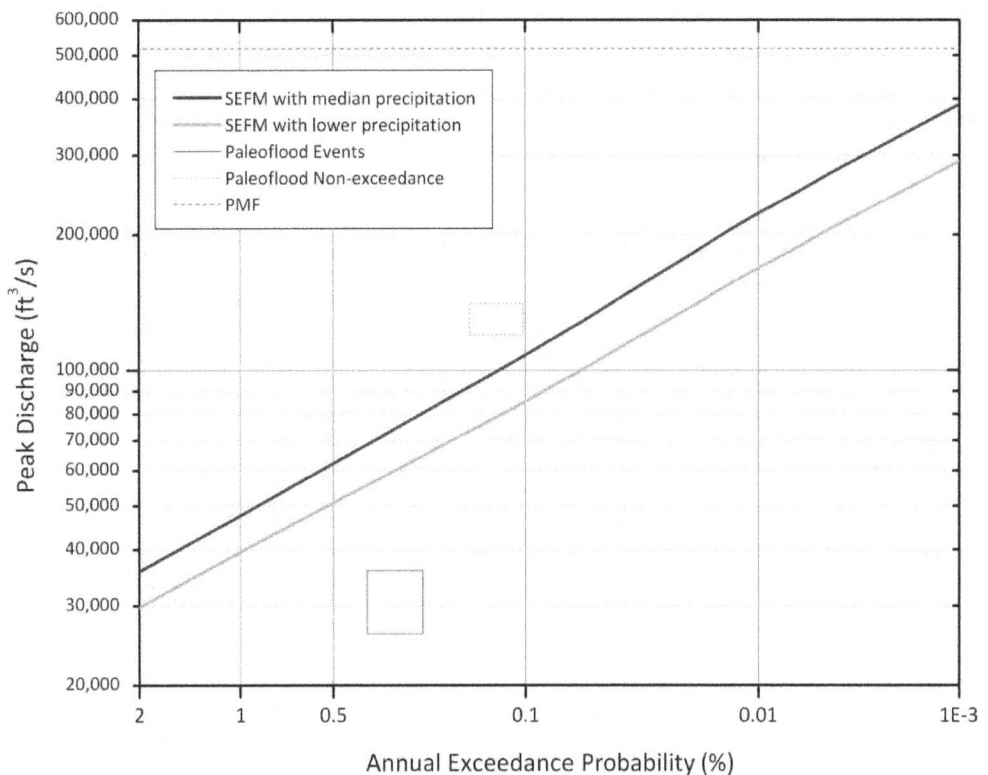

Figure 7–1 Recommended peak discharge frequency curves

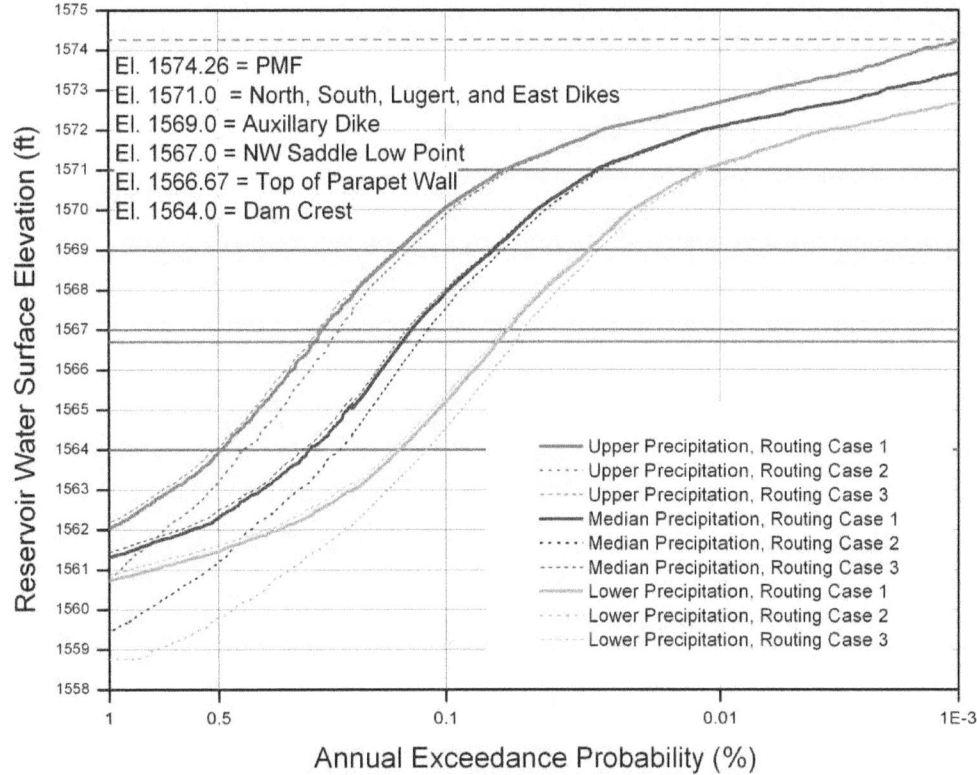

Figure 7–2 SEFM reservoir water surface elevation frequency curves

References

Cohn, T.A., W.L. Lane, and W.G. Baier, 1997, "An algorithm for computing moments-based flood quantile estimates when historical flood information is available," *Water Resources Research* 33(9):2089–2096, http://www.timcohn.com/Publications/CohnLaneBaier97WR01640.pdf.

England, J.F., Jr., R.D. Jarrett, and J.D. Salas, 2003, "Data-based comparisons of moments estimators using historical and paleoflood data," *Journal of Hydrology* 278(4):172–196, http://www.engr.colostate.edu/~jsalas/pdf%20files/58.%20england-jarrett-salas-jhydrol-278.pdf.

Godaire, J.E. and T. Bauer, 2012, "Paleoflood Study, North Fork Red River Basin near Altus Dam, Oklahoma." Technical Memorandum No. 86-68330-2012-14, U.S. Department of the Interior, Bureau of Reclamation, Denver, CO.

MGS Engineering Consultants, Inc. ("MGS Engineering"), 2009, "General Storm Stochastic Event Flood Model (SEFM) - Technical Support Manual, Version 3.70," prepared for the U.S. Department of the Interior, Bureau of Reclamation, Flood Hydrology and Consequences Group, Denver, CO.

Novembre, N.J., V.L. Sankovich, R.J. Caldwell, and J.P. Niehaus, 2012, "Altus Dam Hydrologic Hazard and Reservoir Routing for Corrective Action Study," U.S. Department of the Interior, Bureau of Reclamation Technical Service Center, August 2012, Denver, CO.

Schaefer, M.G., and B.L. Barker, 2002, "Stochastic Event Flood Model," Chapter 20, *Mathematical Models of Small Watershed Hydrology and Applications*, V.P. Singh and D.K. Frevert (eds.), Water Resources Publications, Littleton, CO.

7.6.4 Paleoflood Studies and their Application to Reclamation Dam Safety

Ralph E. Klinger[1]

[1]Bureau of Reclamation, Technical Service Center, P.O. Box 25007, 86-68330,
Denver, CO 80225

The Bureau of Reclamation manages more than 350 water storage dams in the western United States that primarily provide water for irrigation. Periodically, each structure is subjected to a Comprehensive Review (CR) and the safety of each dam is evaluated on the basis of its current condition, past performance, and its projected response to loads that are reasonably expected to test the structural integrity of the structure and its various components. The findings of the CR are used to assess the potential failure risks and potential for loss of life and property. This risk profile allows the Reclamation Dam Safety Office to better manage its available resources to reduce risk and meet established public safety guidelines.

Hydrologic hazards are one of the hazards used in the CR and in the past Reclamation used traditional deterministic criteria (i.e., the Probable Maximum Flood) for evaluating dam safety. Unfortunately, more than half of the dams in the Reclamation inventory failed to meet these criteria and the costs associated with modifying the dams to meet this standard were considered prohibitive. By utilizing probabilistic flood hazard analyses (PFHA), the Dam Safety Office could better categorize its inventory on the basis of risk and prioritize corrective actions. However, using the historical record of flooding alone in a PFHA proved inadequate for accurately estimating the magnitude of less frequent floods. Because paleoflood data can add information on large-magnitude floods outside the historical period of record, as well as place outliers in the record into temporal context, paleoflood studies have long been recognized as an avenue for advancement in evaluating flood hazard.

Beginning in 1993, the Bureau of Reclamation began incorporating paleoflood data into hydrologic hazard estimates. Paleoflood studies were structured to collect the information to meet immediate needs for screening-level risk assessments. Larger, more detailed studies were developed to reduce uncertainty and provide the Dam Safety Office with the information needed to make critical decisions regarding public safety. As the result of almost 20 years of developing paleoflood information at more than 150 Reclamation projects—in addition to advances made by others in the fields of geochronology, hydraulic modeling, and statistical hydrology—the incorporation of paleoflood information has vastly improved Reclamation's ability to accurately estimate flood hazard for its inventory of dams.

PANEL 7

EXTREME STORM SURGE FOR COASTAL AREAS

Co-Chairs:
Donald Resio, University of North Florida
Ty Wamsley, USACE

Technical Reporters:
Mark Fuhrmann and Fernando Ferrante, NRC

8. Extreme Storm Surge for Coastal Areas

Donald Resio[1], Ty Wamsley[2], Mark Fuhrmann[3], Wendy Reed[3], and Fernando Ferrante[4]

[1]University of North Florida, Jacksonville, FL
[2]U.S. Army Corps of Engineers, Vicksburg, MS
[3]Office of Nuclear Regulatory Research, U.S. NRC, Rockville, MD
[4]Office of Nuclear Reactor Regulation, U.S. NRC, Rockville, MD

8.1   Motivation

Storm surge can cause much devastation along coastal areas. Two recent examples are Hurricane Katrina (2005) and Superstorm Sandy (2012). In the case of Hurricane Katrina, almost 1500 people lost their lives, many of those directly or indirectly as a result of storm surge. Although Superstorm Sandy remained at the weakest level of hurricane, its sheer size created an enormous surge of ocean waters, which wiped out many communities and caused billions of dollars in damage. Being able to model and establish the frequency of storm surges is important with regards to design and construction of coastal facilities and for emergency planning purposes.

8.2   Background

Storm Surge is an abnormal rise of water generated by a storm's winds. Storm surge can reach heights well over 20 feet and can span hundreds of miles of coastline. Panel 7 focused on extreme storm surge for coastal areas caused by hurricanes, extratropical cyclones, and intense winter storms. The panel also briefly discussed seiche flooding on closed or semi-closed water bodies. Modeling coastal surge in complex regions requires an accurate definition of the physical system and inclusion of all significant flow processes. Processes that affect storm surge inundation include atmospheric pressure, winds, air-sea momentum transfer, waves, river flows, tides, friction, and morphologic change. A physics-based modeling capability must necessarily link the simulation of winds, water levels, waves, currents, sediment transport, and coastal response (erosion, accretion, and breaching) during extreme events in an integrated system. Numerical models now exist that can properly define the physical system and include an appropriate nonlinear coupling of the relevant processes.

8.3   Overview of Presentations

Joost Beckers presented the Dutch approach to coastal flood hazards assessment. He explained how the Netherlands' response to the 1953 flood led to improvements in flood protection, including development of safety standards in terms of probabilities. He also described the Netherlands' 1990 Water Act, which requires that a periodic safety assessment be performed every five years and that probabilistic methods be used. Statistics such as river discharge and soil cohesion feed into a failure model which gives a probability of failure. Different probabilistic models are used for different water systems; these are known as the Hydra family of models. He also touched on the how uncertainties are dealt with in the Hydra model: epistemic uncertainties are treated as random variables.

Stephen Gill explained how NOAA is using storm surge modeling. One such example is the code SLOSH (Sea Lake and Overland Surges from Hurricanes). This is a simple model, which is driven by the storm track radius of hurricane winds and central pressure, and is used for hurricane evacuation studies. Storm surge probabilities and exceedences are calculated using P-surge. Another model that is used is ETSS (Extra-Tropical Storm Surge), which is a finite differencing model that predicts storm surge heights. It does not include tides, waves, riverflow, or overland storm surge. However, the ETSS website allows one to combine the ETSS output with the hydrological data, such as tidal information. Mr. Gill also described work on sea levels, including estimation of global sea level rise.

Tucker Mahoney talked about FEMA's coastal flood hazard analysis. She explained that the modeling framework involves careful construction of an accurate topographic/bathymentric grid, appropriate storm characterization, accurate storm surge modeling, and an efficient return-period analysis which includes the effects of epistemic uncertainty and overland wave analysis. The data is then used in the flood insurance rate map. FEMA maps the U.S. 1 percent annual-chance flood hazard risk in support of the National Flood Insurance Program (NFIP).

Ty Wamsley explained how the USACE uses a modeling system to look at very-low-probability events and flood response. To simulate storm surge, basin to shelf to floodplain domains with sufficient resolution of the physical system are required. All terms contributing to surge need to be considered (e.g., winds, waves, and tides) and the model system must be validated through comparison to historical data. Dr. Wamsley described CSTORM-MS (Coastal Modeling System), which uses design principles from the Earth System Modeling Framework (ESMF). This framework is an open-source tool that is sponsored by the Department of Defense (DOD), National Science Foundation (NSF), National Aeronautics and Space Administration (NASA), and NOAA. If models are ESMF-compliant, they can be linked to one another. The use of JPM-OS (Joint Probability Method with Optimal Sampling) was mentioned in Dr. Wamsley's talk as well as in Tucker Mahoney's presentation. The JPM-OS method is applied due to the paucity of storm data, which necessitates using numerical simulations of synthetic storms and JPMs to estimate probability density functions and cumulative density functions. This approach replaces the old concept of only planning for storms that have occurred in the historical record in a given area with an approach that considers the range of all storms that can occur in this area. Dr. Wamsley provided a synopsis of a simulation approach for very-low-probability storms through application of a state-of-the-art modeling system with good resolution and physics, including inland propagation of the surge.

Jennifer Irish provided an academic perspective of storm surge studies. She began by highlighting the unique aspects of coastal flooding; for example, coasts exist as a highly dynamic landscape; floods are harder to mitigate due to short timeframes; almost always have multiple hazards at the same time. She showed the participants how surge response functions are calculated, which are primarily proportional to landfall location and central pressure and storm size and hurricane forward speed near landfall, and explained that the flood response includes epistemic uncertainty. In talking about uncertainty, Dr. Irish explained how aleatory uncertainties used to be neglected because it was thought that they were too small for the range of annual exceedence probabilities required in those studies. However, it has since been demonstrated that they can be described mathematically.

8.4   Summary of Panel Discussion

Don Resio opened the panel discussion with a question to Joost Beckers regarding the Netherlands being most exposed to sea level rise and asking how does the country deal with this. Dr. Beckers answered that the Netherlands is considering "optimal" sea level rise. While 2 m is possible, he concluded that it is not likely, but should it be necessary to consider such a height, an adaptive strategy may be required; one that can be made to adjust to different conditions.

The other panel members, when asked about sea level rises gave the following responses. Stephen Gill stated that most of the sea-level rise scenarios and studies found in the current literature only extend out in time to 2100. Comprehensive estimates beyond 2100 have only just begun. Soon, however, for planning purposes to include time horizons for the next 100 years from now, these scenarios will need to be updated to extend beyond 2100. NOAA believes that a scenario-based approach has to be used. NOAA is also trying to look at design and development that considers certain threshold elevations, which leads to design criteria. Ty Wamsley answered that designs and facilities need to be adaptive; to provide room for uncertainty in our prediction of future sea levels.

The next question that was posed was, *"What do you see as the technical and administrative shortcomings when considering coastal surges?"*

Tucker Mahoney answered that FEMA is putting a lot of effort into probabilistic risk determinations, but all of this work gets used to just draw a line on a map. She stated that much more could be done with this information and it should be better fed into the emergency side of the agency. Jennifer Irish stated that we need to make sure there is a good workforce that understands probabilities and hazards. Ty Wamsley commented that a better understanding of MPI is required.

"Relative to PFHA, what does your organization contribute to the evaluation of coastal storms?"

Tucker Mahoney said that FEMA contributes: 1) digital elevation models, 2) synthetic storm suites, and 3) documentation of probabilistic evaluations. Jennifer Irish stated that, in the case of Virginia Tech: 1) Work on three storm types—hurricanes, combined/hybrid storms and Nor'easters—should be extended; 2) studies on how land cover changes impacts of storm effects have been carried out (and are ongoing); and 3) impacts of thresholds on storm effects need to be studied, e.g., the loss of a barrier island that leads to coupled physical and human impacts. Ty Wamsley for USACE: 1) The code CSTORM (Coastal Storm Modeling System) is intended to be open source with the idea of community development and 2) Probabilistic tools to facilitate analysis. Stephen Gill: NOAA is trying to work cross-agency to optimize spatial and temporal observations. NOAA has contributed to 1) improved shallow-water bathymetry and high-resolution topographic maps which will help identify potential threshold events such as overtopping subway entrances, and 2) better understanding of wave effects over inundated areas during storm surges. Joost Beckers: Sharing information and developing probabilistic frameworks and applying them to other agencies.

Tom Nicholson asked the question, *"How would you do a formalized peer review, like the SSHAC process, for storm surge?"*

Don Resio answered that people with alternative opinions and fields of expertise are needed. Scientists and engineers working on a problem do tend to coalesce in their opinion. He stated that most agencies have some level of approach to uncertainties and this needs to be advanced.

Eric Geist from USGS commented on the need to consider joint probabilities in tsunami analysis. He asked if the Poisson function mentioned in Tucker Mahoney's talk is used in storm or hurricane analysis, and if so, is the function used in ensemble models treating uncertainty. Don Resio replied that yes, there is clustering of hurricane tracks and this will be accounted for in the future. Jennifer Irish stated that decadal-scale climate variation was looked at, but that there is no real significant connection, although the baseline may change.

Vincent Rebour from the IRSN in France asked the panel how they thought paleoflood data and the integration of regional information into statistical analyses could be of use in looking at storm surge. Tucker Mahoney stated that good observations are limited to storms since the 1940s and 1950s in general. Data older than that is limited in its quality and reliability. Incorporating data from a wider geographical area adds more uncertainty to the analysis as you may be adding data from an area that is not climatologically similar to the area of interest. Joost Beckers stated that there are a few studies looking at dune erosion going into the 19th century but that the Netherlands has not used them in their analysis. It is hard to get data on the water level and extent of these historical storms. Tucker Mahoney added that looking at paleoflood evidence of storm surges would be a help and that a systematic assessment of palefloods in coastal areas is needed. Don Resio said that this is an under-researched area. It is difficult because such big changes take place in coastal characteristics, such as the loss of marshes, but for those interested in large but low-probability events, such as the NRC, this is something that should be pursued.

"Can an engineering safety factor be used for storm surge protection?"

Jennifer Irish answered that you need to know how things fail for that. You could use engineered failure modes so that you know how they will fail. Don Resio added that an ultraconservative approach could be used, but you need to base this on quantitative analysis to inform the engineering criteria.

8.5   Observations and Insights

*   Significant progress has been made in the past 10 years regarding probabilistic analysis in storm surges.

*   Peer review is very important; Federal agencies should make sure to communicate and coordinate with one another.

*   Education in probabilistic techniques is needed. Some schools only teach deterministic concepts in engineering.

*   Federal agencies are cooperating well in the field of storm surge analysis but there is need for improvement.

*   The use of risk-informed decisionmaking can be expanded and information relating to this should be made more readily available.

8.6   Abstracts

The workshop organizing committee developed the flooding topics to be discussed and chose the panel co-chairs for each panel topic. The co-chairs then identified potential speakers and discussed them with the organizing committee. Following agreement, the co-chairs sent out invitations to the presenters requesting presentation titles and abstracts for documentation in the workshop program. The speakers and panelists are identified in the workshop agenda (please see Appendix A). The following four abstracts document these presentations and, in some cases, reflect the discussions during the panel session.

8.6.1 Coastal Flood Hazard in the Netherlands

Joost Beckers[1], Kathryn Roscoe[1], and Ferdinand Diermanse[1]

[1]Deltares, Delft, the Netherlands

The Netherlands has a long history of fighting floods. About 60% of the country is flood-prone and must be protected by levees or dikes. Up to the first half of the 20th century, design dike heights were roughly 1 m above the highest measured water level. In 1953, a catastrophic storm surge flooded the southern part of the Netherlands. More than 1800 people drowned. In response to this tragedy, a national committee was installed who proposed to shorten the coastline by building storm surge barriers (the Delta Works) and to further decrease the flood hazard by implementing stringent standards for flood protection (Delta Committee, 1958).

Nowadays, the Dutch flood defenses along the major water systems (rivers, lakes, sea) are comprised of a system of dikes (about 3300 km), dunes (about 300 km), storm surge barriers (five) and hundreds of hydraulic structures like sluices. These flood defenses surround about 100 areas called dike rings. Each dike ring has been assigned a safety standard, expressed in terms of a maximum failure probability or return period. This can be translated into a maximum hydraulic load that a structure must be able to withstand.

The hydraulic load on a structure can be a combination of a water level, wave conditions, or other conditions, depending on the failure mechanism under consideration. The maximum hydraulic load for a given return period is determined by a probabilistic computation that takes into account many possible combinations of external forcings that all lead to hydraulic loads on the flood defense. The probabilistic methods are based on statistical analyses of hydrologic and meteorological measurements and on numerical hydrodynamic and wave growth models.

The safety assessment procedure will be illustrated by considering the wave overtopping hazard for a coastal levee. The probabilistic model that is used evaluates the overtopping discharge and probability of occurrence for hundreds of combinations of storm surge and wind-induced waves. The model uses water level and wind speed statistics that are based on observations and that have been extrapolated to exceedance probabilities of 10^{-5} per year. A non-parametric method is used to account for the dependency between wind and water level (De Haan and Resnick, 1977). The SWAN wave growth model (Booij et al., 1999) is used to translate the offshore wind and water level to nearshore wave conditions.

References

Booij, N., R.C. Ris, and L.H. Holthuijsen (1999), "A third generation wave model for coastal regions, Part 1. Model description and validation," *Journal of Geophysical Research* 104(C4):7649–7666, http://cisweb1.ucsd.edu/~falk/modeling/swan1_jgr99.pdf. (SWAN website: http://www.swan.tudelft.nl)

De Haan, L., and S.I. Resnick (1977), "Limit theory for multivariate sample extremes," *Zeitschrift für Warscheinlichkeittheorie und verwandte Gebiete* 40(4):317–337.

Diermanse, F.L.M., and C.P.M. Geerse (2012), "Correlation models in flood risk analysis," *Reliability Engineering and System Safety* 105:64–72, http://dx.doi.org/10.1016/j.ress.2011.12.004.

Roscoe, K., S. Caires, F. Diermanse & J. Groeneweg (2010), "Extreme offshore wave statistics in the North Sea," *WIT Transactions on Ecology and the Environment* 133, http://library.witpress.com/pages/PaperInfo.asp?PaperID=21019.

8.6.2 FEMA's Coastal Flood Hazard Analyses in the Atlantic Ocean and Gulf of Mexico

Tucker B. Mahoney

FEMA – Region IV, Mitigation Division, Risk Analysis Branch, Atlanta, GA 30341

One of the missions of the Federal Emergency Management Agency (FEMA) is to identify and map the nation's 1 percentage-annual-chance flood hazard risk, in support of the National Flood Insurance Program (NFIP). Presently, FEMA is undertaking a substantial effort to update the Flood Insurance Studies (FIS) and Flood Insurance Rate Maps for the populated coastal areas in the United States. These areas are among the most densely populated and economically important areas in the nation.

Coastal flooding on the Atlantic Ocean and Gulf of Mexico coasts is a product of combined offshore, nearshore, and shoreline processes. The interrelationships of these processes are complex and their relative effects can vary significantly by location. These complexities present challenges in mapping a local or regional hazard from coastal flooding sources. Congress' mandate to FEMA to assess and map the 1 %-annual-chance flood hazard based on present conditions simplifies some of these complexities. As part of the study process, FEMA also identifies the 0.2 %-annual-chance flood hazard for coastal areas. In some cases, higher-frequency risks, such as the 2 %, 10 %, 25 % and 50 % hazards, will be examined and made available to communities for informational purposes.

The most severe storms, which dominate the coastal 1 %-annual-chance flood risk, on the Atlantic Ocean and Gulf of Mexico coast can generally be classified as either tropical systems or northeasters. Because long-term records of water level are spare compared to the risk, FEMA's Guidelines and Specifications require that a probabilistic assessment of storm surge and coastal flooding be undertaken where applicable (FEMA, 2007). Ongoing coastal flood risk studies utilize the Joint Probability Method-Optimal Sampling (JPM-OS) to statistically model the spatial and temporal occurrence and characteristics of tropical systems, typically hurricanes (Resio, 2007, and Toro et al., 2010). For the mid- to north-Atlantic coastline, northeasters also contribute to the 1 %-annual-chance coastal flood risk. The contribution of these storms to the flood hazard risk is evaluated using the Empirical Simulation Technique (Scheffner et al., 1999).

For each individual storm to be evaluated (be it historical or synthetic), the still-water elevation, including wave setup, is determined using a two-dimensional hydrodynamic model and a two-dimensional wave model. Recent FEMA coastal flood hazard studies utilize the coupled SWAN+ADCIRC model to couple together the still-water level and wave modeling (Dietrich et al., 2012). The model mesh typically extends well inland from the coast in order to capture the full extent of the storm surge which propagates overland during an extreme event. The smallest cell spacing is typically between 50 and 100 feet and, depending on location, the mesh may extend to the 30- or 50-foot NAVD88 contour. Results of these simulations are used to statistically determine the still-water elevation, including wave setup, for a given recurrence interval (e.g., a 1 % annual chance).

After determining the statistical still-water elevation, the next step is to simulate the propagation of waves overland using the one-dimensional model Wave Hazard Analysis for Flood Insurance Studies (WHAFIS), which was developed by the National Academy of Sciences for this purpose (NAS, 1977). Input for WHAFIS is drawn from the near-shore wave model. From the WHAFIS results, FEMA then maps then delineates and maps flood hazard zones on the Flood Insurance Rate Map.

Although a coastal FIS uses state-of-the-art techniques, improvements are still possible and will likely occur in coming years. For example, more research is needed into two-dimensional overland wave modeling to truly characterize wave conditions and interactions at a specific location. It would be beneficial if variable future conditions (from factors such as shoreline change, sea-level rise, and/or changes in climate) and their impact on the flood hazard could be accounted for. Lastly, a full uncertainty and error analysis is needed.

Other federal agencies and scientific bodies may benefit from some of the tools produced by FEMA for an updated coastal FIS. For each study, a detailed seamless topographic and bathymetric digital elevation model will be created for most locations in the coastal United States using the best and most recently available data. A highly detailed, fully attributed numerical model mesh will be created for most locations on the Atlantic Ocean and Gulf of Mexico coasts and may also be utilized for other flood hazard studies. A mesh can be easily modified to add detail to an area of interest or expanded further upland to simulate larger storm surges than those encountered during a FEMA coastal flood hazard study. Lastly, the JPM framework and variety of input storm conditions may provide useful information to future assessments of coastal flood risk with lower probabilities of occurrence.

References

Dietrich, J.C. et al. (2012), "Performance of the Unstructured-Mesh SWAN+ADCIRC Model in Computing Hurricane Waves and Surge," *Journal of Scientific Computing* 52(2):468–497, http://www3.nd.edu/~coast/reports_papers/2012-JSC-dtwdlzhsww-SWAN+ADCIRC-Performance-INPRESS.pdf.

Federal Emergency Management Agency (FEMA, 2007), "Atlantic Ocean and Gulf of Mexico Coastal Guidelines Update," Final Draft, Washington, DC, http://www.fema.gov/library/viewRecord.do?id=2458.

National Academy of Sciences (NAS, 1977), "Methodology for Calculating Wave Action Effects Associated with Storm Surges", Washington, DC.

Resio, D.T., 2007, "White Paper on Estimating Hurricane Inundation Probabilities." Version 11, Engineer Research and Development Center (ERDC), US Army Corps of Engineers, Vicksburg, MS.

Scheffner, N.W., et al. (1999), "Use and Application of the Empirical Simulation Technique: User's Guide," Technical Report CHL-99-21, U.S. Army Corps of Engineers' Engineer Research Development Center, Vicksburg, MS, http://www.dtic.mil/cgi-bin/GetTRDoc?AD=ADA376132.

Toro, G., et al. (2010), "Quadrature-based approach for the efficient evaluation of surge hazard," *Ocean Engineering* 37(1):114–124, http://dx.doi.org/10.1016/j.oceaneng.2009.09.005.

8.6.3 Modeling System for Applications to Very-Low Probability Events and Flood Response

Ty V. Wamsley and T. Christopher Massey

U.S. Army Corps of Engineers, Coastal and Hydraulics Laboratory (CHL), Vicksburg, MS 39183

Modeling coastal surge in complex regions requires an accurate definition of the physical system and inclusion of all significant flow processes. Processes that affect storm surge inundation include atmospheric pressure, winds, air-sea momentum transfer, waves, river flows, tides, friction, and morphologic change. Numerical models now exist that can properly define the physical system and include an appropriate nonlinear coupling of the relevant processes. A coupled system of wind, wave, circulation, and morphology models has been developed and implemented for regions in the Gulf of Mexico, Atlantic, and Great Lakes. The U.S. Army Corps of Engineers (USACE) Engineer and Research Development Center's (ERDC's) Coastal Storm Modeling System (CSTORM-MS) (Massey et al., 2011) includes a tropical planetary boundary layer model, TC96 MORPHOS-PBL (Thompson and Cardone, 1996), to generate the cyclone wind and pressure fields. It is also possible to use a variety of other wind products including both measured and hindcast winds. The storm surge and current fields are modeled with the ocean hydrodynamic model ADCIRC (Luettich et al., 1992) which computes the pressure- and wind-driven surge component. The regional and nearshore ocean-wave models, WAM (Komen et al 1994) and STWAVE (Smith et al 2001) generate the wave fields. The AdH model (Berger and Howington 2002) is applied to simulate the nearshore zone where morphology change is computed and utilizes the CHL 2-Dimensional nearSHORE (C2SHORE) sediment transport and morphology change code through a call to its SEDiment transport LIBrary (SEDLIB).

A physics-based modeling capability must necessarily link the simulation of winds, water levels, waves, currents, sediment transport, and coastal response (erosion, accretion, and breaching) during extreme events in an integrated system. A flexible and expandable computational coupler (CSTORM coupler) uses design principles from the Earth System Modeling Framework (ESMF) and links the atmosphere, circulation, waves, and morphology models for application on desktop computers or high-performance computing resources. The computational coupler employs a two-way coupling scheme that not only enhances the represented physics, but also results in significant improvements in computational time when compared to file-based approaches. A series of integrated graphical user interfaces within the Surface-water Modeling System (SMS) allows for convenient setup and execution of these coupled models.

The ESMF is a set of open-source software tools for both building and linking complex weather, climate, and related models. It has support from many agencies within the Department of Defense, the National Oceanic and Atmospheric Administration (NOAA), and the National Aeronautics and Space Administration (NASA). In order for models to use ESMF, they need to be ESMF-compliant, which means that their computer code is organized into three distinct phases: initialization, run, and finalization. Models that meet this standard allow for an almost "plug-and-play" capability when being used within an ESMF-enabled coupler such as the CSTORM coupler. Therefore, the system is well suited for expansion to include effects from urban flooding or flood propagation at the facility scale.

Flood response operations and hazard mitigation during an event requires knowledge not only of the ultimate high-water elevation, but also the timing of the flood wave. A reasonable estimation of the flood hydrograph requires the application of a high-resolution model. The modeling system described allows flexible spatial discretizations which enable the computational grid to be created with larger elements in open-ocean regions where less resolution is needed, and smaller elements can be applied in the nearshore and estuary areas where finer resolution is required to resolve hydrodynamic details and more accurately simulate the timing of storm surge propagation. However, application of high-fidelity modeling such as that described above can be time-prohibitive in operational mode. The USACE has developed a proof-of-concept operational flood prediction system that uses pre-computed high-fidelity simulations stored in a database applying surrogate modeling strategies. This approach provides multi-process predictions in seconds for flood hazard assessment and mitigation.

References

Berger, R. C. and S.E. Howington, 2002, "Discrete Fluxes and Mass Balance in Finite Elements," Journal of Hydraulic Engineering 128(1):87–92, http://dx.doi.org/10.1061/(ASCE)0733-9429(2002)128:1(87).

Komen, G.J., L. Cavaleri, M. Donelan, K. Hasselmann, S. Hasselmann, and P.A.E.M. Janssen. 1994, Dynamics and Modelling of Ocean Waves, Cambridge University Press, Cambridge, UK.

Luettich, R.A., J.J. Westerink, and N.W. Scheffner, 1992, "ADCIRC: An Advanced Three-Dimensional Circulation Model for Shelves, Coasts, and Estuaries; Report 1: Theory and Methodology of ADCIRC-2DDI and ADCIRC-3DL," Technical Report DRP-92-6, U.S. Army Corps of Engineers, Washington, DC, http://www.dtic.mil/dtic/tr/fulltext/u2/a261608.pdf.

Massey, T.C., T.V. Wamsley, and M.A. Cialone, 2011, "Coastal Storm Modeling-System Integration," Proceedings of Solutions to Coastal Disasters 2011, American Society of Civil Engineers, http://dx.doi.org/10.1061/41185(417)10.

Smith, J.M., A.R. Sherlock, and D.T. Resio, 2001, "STWAVE: Steady-State Spectral Wave Model User's Manual for STWAVE, Version 3.0," ERDC/CHL SR-01-1, U.S. Army Corps of Engineers Engineer Research and Development Center, Vicksburg, MS, http://chl.erdc.usace.army.mil/Media/2/4/4/erdc-chl-sr-01-11.pdf

Thompson, E.F., and V. J. Cardone, 1996, "Practical Modeling of Hurricane Surface Wind Fields," Journal of Waterway, Port, Coastal and Ocean Engineering 122(4):195–205, http://champs.cecs.ucf.edu/Library/Journal_Articles/pdfs/Practical%20modeling%20of%20hurric ane%20surface%20wind%20fields.pdf

8.6.4 Coastal Inundation Risk Assessment

Jennifer L. Irish

Department of Civil and Environmental Engineering, Virginia Tech, Blacksburg, VA 24061

In the last decade, the United States has experienced some of its largest surges and hurricane-related damages on record. Understanding the risk posed by extreme storm surge events is important both for effective evacuation and for future coastal engineering planning, management, and design. For the first, effective evacuation in advance of a hurricane strike requires accurate estimation of the hurricane surge hazard that effectively conveys risk not only to government decisionmakers but also to the general public. Two primary challenges exist with the current structure for surge warning. First, existing computational methods for developing accurate, quantitative surge forecasts, namely surge height and inundation estimation, are limited by time and computational resources. Second, due primarily to the popularity and wide use of the Saffir-Simpson wind scale to convey the complete hurricane hazard, the public's perception of surge hazard is inaccurate.

For the second, reliable extreme-value hurricane flooding estimates are essential for effective risk assessment, management, and engineering in the coastal environment. However, both a limited historical record and the range of, and uncertainty in, future climate and sea-level conditions present challenges for assessing future hurricane flooding probability. Historical water-level observations indicate that sea level is rising in most hurricane-prone regions, while historical observations of hurricane meteorology indicate decadal variation in hurricane patterns, including landfall location and rate of occurrence. Recent studies, including those by the Intergovernmental Panel on Climate Change (IPCC), also suggest that in the future sea-level rise may accelerate and major tropical cyclones may intensify.

Methods will be presented for robustly and efficiently quantifying both forecast and extreme-value surge statistics, where the latter will incorporate sea-level rise and time-varying hurricane conditions. A joint probability approach will be used with surge response functions to define continuous probability density functions for hurricane flood elevation. Joint probability is assigned based on the probability of the hurricane meteorological parameters, such as rate of hurricane landfall, central pressure, and storm radius. From a theoretical standpoint, surge response functions are scaling laws derived from high-resolution numerical simulations, here with the hydrodynamic model ADCIRC. Surge response functions allow rapid algebraic surge calculation based on these meteorological parameters while guaranteeing accuracy and detail by incorporating high-resolution computational results in their formulation. Thus the use of surge response functions yields continuous joint probability density functions. The form of the surge response functions allows direct assessment, within the joint probability framework, of alternate probability distributions for any hurricane parameter.

Uncertainty in the probabilistic estimates will be discussed in the context of extreme-value statistics by considering the variability in future climate and sea-level projections. Here, we use dimensionless scaling and hydrodynamics arguments to quantify the influence of hurricane variables and regional geographic characteristics on the surge response.

In summary, physical attributes of hurricane surge can be quantified using forecasted, historical, and hypothetical hurricane track information, and this information may be used to more rapidly and accurately convey surge hazard to planners, decisionmakers, and the public.

Note: Abstract is modified from those presented at the American Geophysical Union's Fall Meetings in 2011 and 2012.

PANEL 8

COMBINED EVENTS FLOODING

Co-Chairs:
David Margo, USACE
Joost Beckers, Deltares

Technical Reporters:
Michelle Bensi and Jeff Mitman, NRC

9.   Combined Events Flooding

David Margo[1], Joost Beckers[2], Michelle Bensi[3], and Jeff Mitman[4]

[1]U.S. Army Corps of Engineers, Pittsburgh, PA
[2]Deltares, the Netherlands
[3]Office of New Reactors, U.S. NRC, Rockville, MD
[4]Office of Nuclear Reactor Regulation, U.S. NRC, Rockville, MD

9.1   Motivation

Site-specific flood hazards can result from various combinations of flood-causing mechanisms. Floods can occur due to a single uncommon event such as the probable maximum precipitation or floods can occur due to an unforeseen combination of events associated with natural (e.g., rainfall), human-made (e.g., dam failure), and organizational (e.g., operating error) factors. Combined events can include flooding caused by seismically induced dam or levee failure; flooding caused by combinations of snowmelt, rainfall, and ice; flooding caused by combinations of coastal and riverine events; basin- or system-wide performance issues; human and organizational factors; and many other combined scenarios. As a result, no single prescriptive set of scenarios is adequate as the basis for the design flood. The design of facilities subjected to flood hazards must consider all of the plausible site-specific flood hazards including those hazards resulting from the uncommon combination of common events. Probabilistic flood hazard assessment can support the identification and evaluation of these combined event scenarios within a risk-informed decisionmaking framework.

9.2   Background

Design-basis flood scenarios have historically been associated with probable maximum events of deterministic origin. Several combined event scenarios are usually derived from various combinations of flood-causing mechanisms because no single flood-causing event is adequate as a design basis for a critical facility. Each of these combined scenarios are formulated to achieve notional annual exceedance probability goals on the order of 10^{-6}; however, the actual probability for these events is unknown and likely varies considerably from one scenario to another and from one site to the next. As a result, the criticality of each scenario is unknown because its contribution to the system risk is unknown. Also, it is unknown whether or not the most critical scenarios have even been identified and considered in the standard set of probable maximum events. It is often the case that more likely combinations resulting in less than probable maximum events pose a greater risk. Probabilistic flood hazard assessment provides a robust framework to tackle these challenges by explicitly estimating and considering the probability of these combined event scenarios within a risk-informed framework.

9.3   Overview of Presentations

Combined Events in External Flooding Evaluation for Nuclear Plant Sites

Kit Ng described some of the current guidance and state of the practice for selecting a design-basis flood event. A deterministic framework is used to identify relevant and plausible combinations of events. These combinations are typically comprised of a primary event such as the probable maximum flood and one or more secondary coincident or antecedent events. Reasonable combinations are obtained by considering potential dependencies and correlations between the events. For example, maximum earthquakes are typically not combined with maximum floods. A notional goal of 1E-6 is generally considered acceptable for the annual probability of occurrence for the event combinations. Although the approach is deterministic, the developers of the guidance recognized the limitations of the methodology and the gaps in knowledge. Some considerations in the transition to a probabilistic or risk-informed methodology include the level of effort required for the assessment, the criteria for acceptability, improvement in the explicit treatment of uncertainties, and the relationship between deterministic and probabilistic approaches.

Assessing Levee System Performance Using Existing and Future Risk-Analysis Tools

Chris Dunn described the risk-analysis tools being used by USACE to conduct systems-based risk assessments. Existing tools such as the Flood Damage Reduction Analysis, HEC-FDA, have been used effectively to evaluate risk for complex systems such as the Sacramento River, but these tools do not explicitly address the problem as an interrelated system. Emerging tools such as the Watershed Analysis Tool, HEC-WAT, with the FRA compute provide a systems and life cycle modeling approach to assess risks and uncertainties in both simple and complex systems. For example, failure of an upstream levee and the resulting loss of hydrograph volume at a downstream location can be explicitly accounted for in the model. The model applies Monte Carlo simulation techniques with new computational methodologies to evaluate risks for the entire system. Some of the technical challenges faced by USACE include the consideration of rehabilitation, repair and flood recovery in the life cycle analysis; tradeoffs between detailed modeling and important sources of uncertainty; and computational burden.

Seismic Risk of Co-Located Critical Infrastructure Facilities – Effects of Correlation and Uncertainty

Martin McCann described a parametric study that was used to illustrate the influence of dependence and correlation on the seismic hazard at a critical facility. It was noted that a common taxonomy for uncertainty is critically important to ensure that key sources of uncertainty are not overlooked (even if they cannot be quantified) and that they are not double-counted. Sources of dependence and correlation included the epistemic uncertainties in the seismic hazard estimates, the aleatory uncertainty in the median ground motions, and the spatial correlation of ground motions for sites located in proximity to one another. Logic trees can be used to model the uncertainties and account for alternative interpretations of the available scientific data. Portraying probabilistic seismic hazard analysis results as a set of discrete curves for each alternative interpretation can provide additional insights over a portrayal of the median hazard curve and its associated confidence bands. Other important factors to consider include that fact the median estimate of ground motions can systematically, but randomly, vary from one earthquake to the next. The evaluation of risk is site- and scenario-specific and may depend on the number of affected facilities, the location of the facilities relative to each other, the seismic fragility of the facilities, the location of seismic sources relative to the facilities, and the magnitude of earthquakes that can occur. Depending on these factors, the impacts to the risk can range from relatively small (factor of about 2) to considerably greater (factor of 10 or more).

Storm Surge – Riverine Combined Events

Joost Beckers described a probabilistic flood hazard assessment for a tidal river system near the city of Rotterdam. The analysis considered relevant combinations of forcing variables such as river discharge, sea level, wind speed, wind direction, and storm surge barrier operation. The Sobek one-dimensional hydraulic model was used to obtain peak water levels at various locations within the system for each combination. Approximately 7000 combinations were evaluated, which resulted in a significant computational burden. Probability estimates for each combination were made and integrated across all combinations to obtain a water level for a given probability at key locations in the system. These water levels could then be compared to the height of the levees to evaluate their adequacy. Results showed that large contributions to risk were obtained from combinations of storm surges and discharges that were much less than an extreme event. Human factors associated with operation of the storm surge barrier were also a significant contributor to the risk. Several challenges were identified which include computational burden for the large number of combinations, proper accounting for correlations, and the need for a site-specific model.

Combining Flood Risks from Snowmelt, Rain, and Ice – The Platte River in Nebraska

Doug Clemetson described a flood frequency study for a wide, shallow, and braided river that is susceptible to ice formation during cold winters. When ice is present, it can have a significant impact on flood stages. Floods in this system can occur due to spring snowmelt, ice effects during the spring melt, and summer rainfall events. The analysis initially used the methodology from FEMA Publication 37 for analysis of ice-jam flooding. Use of this method resulted in the overestimation of flood risk because of the fact that the river is not affected by ice every year. A new methodology was developed to properly account for the fraction of years in which peak stages are affected by ice. This methodology provides a simple and practical approach to solve a complex problem. The new methodology has been adopted by FEMA and has been published in Appendix F of the Guideline and Specifications for Flood Hazard Mapping Partners.

Human, Organizational, and Other Factors Contributing to Dam Failure

Pat Regan described several dam failures that occurred as a result of combinations of events that are much more likely to occur than extreme design-basis events. Failures are seldom the result of a single root cause. Most failures result from an unforeseen combination of events that include natural events, human-made events, and organizational factors. These events can be known if an effort is made to recognize them. In hindsight, we often realize what we could have known. If too much emphasis is placed on evaluating extreme events, the more likely system failures will be missed. A dam failure case history was presented as an example of a system failure. Some of the contributors to the failure included a significant precipitation event, a mechanical fault that prevented operation of the spillway, flooded roads that hindered access to the site, fallen transmission poles that resulted in the loss of communication, occurrence on a weekend that made it difficult to mobilize equipment and operators, occurrence at night that prevented mobilization of a helicopter to access the site, and several other factors. This combination of events led to overtopping and subsequent failure of the embankment dam. Failure of the South Fork Dam was presented as another case history. The dam was poorly constructed and the spillway was inadequate due to a combination of crest lowering, settlement, and installation of fish screens. The dam overtopped and failed during a flood in 1889 resulting in more than 2200 deaths.

9.4   Summary of Panel Discussion

Panel Questions:

    i.   *How can a risk-informed framework be used to identify plausible event combinations that are relevant to the flood hazard assessment?*

    ii.   *How can we estimate the probabilities and risks associated with event combinations?*

Over time, the technical community has perhaps become overly quantitative, thus losing sight of the big picture. Groupthink occurs at times and too much emphasis has been placed on meeting the regulatory requirement instead of thinking about what events could happen and the implications of those events if they were to happen. Traditionally the dam safety community has assessed the adequacy of a dam based on three questions: "Is the dam standing today?", "Will the dam withstand a large flood?", and "Will the dam withstand a large earthquake?". Placing too much emphasis on this limited set of prescriptive design-basis events has perhaps stymied the imagination of practitioners. Case histories reveal that most failures occur due to a combination of natural events, human-made events, and organizational factors. A paradigm shift toward a more comprehensive systems approach will facilitate better risk-informed analysis and decisions. While we still need to account for the extreme design events, we cannot overlook the combination of more frequent events which can be more likely to occur and can pose a bigger threat. Risk-informed methodologies should follow a systems approach, including the consideration of relevant event combinations including the uncommon combination of not-uncommon events. This requires a transition from a design perspective to a prediction perspective. A risk-informed framework supports the explicit consideration of all relevant event combinations.

Given adequate resources, the technical community has the capacity to identify relevant combinations of events that contribute to risk. Factors such as policies, politics, funding, and other non-technical issues often play a more critical role in determining the quantity and quality of information that can be developed for decisionmakers. The technical analysis often requires significant computational burden to address these complex problems. A significant challenge is being able to calibrate and run models efficiently to obtain believable probability estimates for very rare events. How does one validate a logic tree with 60,000,000 end branches? How does one efficiently implement a model that requires days of run time? In our quest for the truth, we are very good at making many assumptions that allow us to precisely calculate things that are very uncertain. Models that can solve complex problems are certainly needed. What is also needed are methods and guidance for making credible and practical simplifications within these models so that analyses can be accomplished efficiently and results can be communicated effectively.

Case histories provide an opportunity to learn from the past. Documenting and sharing successes, failures, and near misses can provide valuable insight to a risk assessment. Increased sharing of information within industries and across industries is needed. The earthquake community of practice provides a good model for capturing information and lessons learned from both successes and failures.

Emergency exercises provide an opportunity to prepare for the future. Evaluating performance of the system under simulated emergency conditions can reveal unforeseen vulnerabilities. Operation and maintenance practices impact performance and risk. Issues related to site access during a flood, condition of the mechanical and electrical equipment, availability of backup systems, staffing levels, staff qualifications, and others should all be considered when assessing risk.

9.5   Abstracts

The workshop organizing committee developed the flooding topics to be discussed and chose the panel co-chairs for each panel topic. The co-chairs then identified potential speakers and discussed them with the organizing committee. Following agreement, the co-chairs sent out invitations to the presenters requesting presentation titles and abstracts for documentation in the workshop program. The speakers and panelists are identified in the workshop agenda (please see Appendix A). The following six abstracts document these presentations and, in some cases, reflect the discussions during the panel session.

9.5.1 Combined Events in External Flooding Evaluation for Nuclear Plant Sites

Kit Y. Ng[1]

[1]Bechtel Power Corporation, Geotechnical & Hydraulic Engineering Services, Frederick, MD 21703

The presentation will review the historical and current practices of assessing combined events for flooding caused by external events at nuclear plant sites. The current industry and regulatory guides on the evaluation of combined events flooding, including RG 1.59, NUREG/CR-7046, NUREG-0800, and ANSI/ANS 2.8-1992 are primarily deterministically based. As stated in ANS 2.8, no single flood-causing event is an adequate design base for a power reactor. Combined events can be sequential or coincidental occurrences of attendant or causative flood causes. The criteria for selecting the set of combined events depend on the location of the plant and the characteristics of the flood-causing phenomena specific to the site. ANS 2.8 also indicates that an average annual exceedance probability less than 1×10^{-6} is an acceptable goal for selection of flood design bases for power reactor plants. This paper examines the fundamentals of combined events in the deterministic framework. In addition, it explores the challenges of evaluating external flooding in the probabilistic space.

9.5.2 Assessing Levee System Performance Using Existing and Future Risk Analysis Tools

Christopher N. Dunn, P.E., D.WRE[1]

[1]USACE Hydrologic Engineering Center, Davis, CA

A process was defined to apply risk-analysis methodologies to identify potential system-wide hydraulic impacts resulting from modifications to the Sacramento River Flood Control Project (SRFCP). This effort demonstrated that existing risk-analysis tools can be applied in a systems context to reveal responses of one region of a system to perturbations from another region. The example application illustrates the complexities and effort required to conduct a system-wide risk analysis. U.S. Army Corps of Engineers (USACE) policy, as stated in ER 1105-2-101, "Risk Analysis for Flood Damage Reduction Studies" (USACE, 2006a), requires the use of risk analysis and its results in planning flood risk-management studies and these are to be documented in principal decision documents. The goal of the policy is a comprehensive approach in which the key variables, parameters, and components of flood risk-management studies are subject to probabilistic analysis. The benefit of the process for the evaluation of proposed modifications to the SRFCP is an increased understanding of the potential risk inherent in modification alternatives. A second, but no less important, goal of this exercise was to understand more fully what is required to advance the current methods and tools for risk-management assessments. Thus, a major purpose of this effort was also to identify and assist the development of methods, tools, and guidance for performing and using risk and reliability assessments that match the complexity and frequency of the assessments.

To address these needs, the U.S. Army Corps of Engineers Institute for Water Resources' Hydrologic Engineering Center (CEIWR-HEC) has developed the next generation of flood risk-analysis tool, HEC-WAT/FRA (Watershed Analysis Tool with the Flood Risk Analysis option). HEC-WAT streamlines and integrates a water-resources study using software commonly applied by multidisciplinary teams. Software such as HEC-HMS, HEC-SSP, HEC-RAS, HEC-ResSim, HEC-DSSVue, and HEC-FIA are currently implemented within HEC-WAT, allowing a study team to perform many of the necessary hydrologic, hydraulic, and planning analyses all orchestrated through a single interface. HEC-WAT allows a Project Delivery Team (PDT) to perform a risk analysis within a systems context in an intuitive and collaborative manner.

HEC-WAT provides the framework to coordinate the study, while the individual pieces of software provide the analytical computations. The central framework also allows the user to: load GIS-based layers; establish stream networks and schematics; identify locations where models would share information; define the modeling programs and their sequence order; import and edit existing models; develop new models; organize and store data; organize and develop alternatives, analysis periods, and simulations; run modeling programs directly; and view and compare alternative results.

While HEC-WAT provides tools to help address many of the issues facing a Corps study, the requirement for a life-cycle analysis is not addressed. Corps policy states that a life-cycle analysis will be performed while still performing a risk analysis that allows for parameter sampling across all models. This requirement has existed for over two decades yet there are still few tools and little guidance to support this requirement. For this reason, CEIWR-HEC developed an option within HEC-WAT that will analyze complex riverine systems while addressing the flood risk-management, systems, and life-cycle requirements. This new option, FRA (Flood Risk Analysis), is a computation option from the HEC-WAT software (Figure 9–1) that allows a user to perform plan formulation or system-performance analyses while incorporating risk and life-cycle capabilities.

HEC-WAT with FRA provides many capabilities: systems approach; event-based sampling; ability to do scenario analysis; and, structure-by-structure, cost, non-structural, loss-of-life, and agricultural damage analysis. Addition of FRA to HEC-WAT accommodates many, if not all, of the recommendations that the Corps agreed with from the National Research Council (NRC, 2000) report on the Corps' implementation of risk analysis for flood damage reduction. It will also aid in implementing the Chief of Engineers' Campaign Goals.

The FRA option includes sampling and solution techniques, uncertainty definitions, and system-wide component fragility and performance interactions/relationships for these complex riverine systems. The capabilities evolved from previous efforts and as detailed in the software design document (USACE, 2008), which proposes significant advancements to the USACE modeling approach for risk assessments. When implemented, the new feature could be used nationwide for levee certification, levee assessment, planning, and design studies and advance the Corps modeling approach for risk and life-cycle analysis.

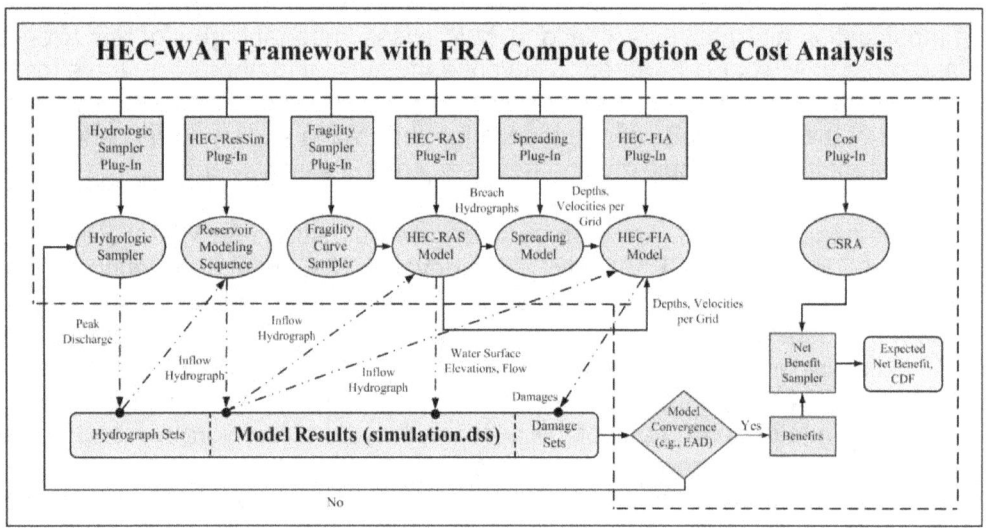

Figure 9–1 HEC-WAT Framework with FRA Compute Option

To test the many features and capabilities of HEC-WAT/FRA, the software has been applied on the Columbia River as part of the Columbia River Treaty (CRT) 2014/2024 review. The CRT review is a regional effort within the Corps of Engineers' Northwestern Division (CENWD), requiring close coordination amongst the Division headquarters and the three Districts with geographic responsibilities within the Columbia River Basin: Portland, Seattle, and Walla Walla. Part of the review is to develop a comprehensive systems approach for the Columbia River Basin and to evaluate the current and future flood risk within the Basin. The purpose of the technical studies is to document the base condition for the CRT 2014 review. HEC-WAT with the flood risk (FRA) option is being used to evaluate treaty alternatives using EAD and other criteria.

References

Dunn, C.N. and P.R. Baker, 2010, "A Watershed Modeling Tool, HEC-WAT", *Watershed Management 2010: Innovations in Watershed Management Under Land Use and Climate Change*, American Society of Civil Engineers, Madison, WI, http://dx.doi.org/10.1061/41143(394)99.

U.S. Army Corps of Engineers (USACE), 2009, "Documentation and Demonstration of a Process for Risk Analysis of Proposed Modifications to the Sacramento River Flood Control Project (SRFCP)," Project Report 71, Davis, CA, http://www.hec.usace.army.mil/publications/ProjectReports/PR-71.pdf.

National Research Council, 2000, "Risk Analysis and Uncertainty in Flood Damage Reduction Studies," National Academies Press, Washington, DC.

USACE, 2000, "Planning Guidance Notebook", Engineer Regulation (ER) 1105-2-100, Washington, DC, http://publications.usace.army.mil/publications/eng-regs/ER_1105-2-100/ER_1105-2-100.pdf.

USACE, 2006, "Risk Analysis for Flood Damage Reduction Studies," ER 1105-2-101, Washington, DC, http://planning.usace.army.mil/toolbox/library/ERs/er1105-2-101.pdf.

USACE, 1996, "Risk-Based Analysis for Flood Damage Reduction Studies," Engineer Manual 1110-2-1619, Washington, DC, http://publications.usace.army.mil/publications/eng-manuals/EM_1110-2-1619_sec/EM_1110-2-1619.pdf.

USACE, 2010, "USACE Process for the National Flood Insurance Program (NFIP) Levee System Evaluation," Engineer Circular 1110-2-6067, Washington, DC, http://publications.usace.army.mil/publications/eng-circulars/EC_1110-2-6067.pdf.

Dunn, C.N., G.W. Brunner, and D.J. Harris, 2005, "Software Integration for Watershed Studies: Hydrologic Engineering Center's Watershed Analysis Tool (HEC-WAT)," *Proceedings of the World Water and Environmental Resources Congress 2005: Impacts of Global Climate Change*, American Society of Civil Engineers, Anchorage, AK, http://dx.doi.org/10.1061/40792(173)490.

Dunn, C.N., 2006, "The Development of a Software Integration Tool for Watershed Studies - The Hydrologic Engineering Center's Watershed Analysis Tool (HEC-WAT)," *Proceedings of the Third Federal Interagency Hydrologic Modeling Conference*, Reno, NV.

Dunn, C.N., 2007, "Integrated Water Resources Analysis Using the Watershed Analysis Tool (HEC-WAT)," *Proceedings of the World Environmental and Water Resources Congress 2007*, American Society of Civil Engineers, Tampa, FL.

Dunn, C.N. and R.A. Pietrowsky, 2008, "Collaborative Modeling for Decision Making," *Proceedings of the World Environmental and Water Resources Congress 2008: Ahupua'a*, American Society of Civil Engineers, Honolulu, HI, http://dx.doi.org/10.1061/40976(316)522.

Dunn, C.N., and M.K. Deering, 2009, "Flood Risk Assessment of Complex Riverine Systems," *Proceedings of the World Environmental and Water Resources Congress 2009: Great Rivers*, American Society of Civil Engineers, Kansas City, MO, http://dx.doi.org/10.1061/41036(342)341.

Harris, D.J., C.N. Dunn, and M.K. Deering, 2010, "Flood Risk Assessment of Complex Riverine Systems," *Proceedings of the World Environmental and Water Resources Congress 2010: Challenges of Change*, American Society of Civil Engineers, Providence, RI, http://dx.doi.org/10.1061/41114(371)483.

Dunn, C.N., P.R. Baker, and B.A. Faber, 2011, "Levee System Performance Using Future Risk Analysis Tools," *Proceedings of the World Environmental & Water Resources Congress 2011: Bearing Knowledge for Sustainability*, American Society of Civil Engineers, Palm Springs, CA, http://dx.doi.org/10.1061/41173(414)241.

Dunn, C.N., and P.R. Baker, 2013, "A Nationwide Comprehensive Water Management System: Can We Get There and Why Should We Care" (draft), *Proceedings of the World Environmental and Water Resources Congress 2013: Showcasing the Future*, American Society of Civil Engineers, Cincinnati, OH, http://dx.doi.org/10.1061/9780784412947.200

Hydrologic Engineering Center, (2008). "*Model Design for Flood Damage and Risk Analysis of Complex Riverine Systems*". USACE, Institute for Water Resources, 609 Second Street, Davis, CA. 47 pages.

9.5.3 Seismic Risk of Co-Located Critical Infrastructure Facilities – Effects of Correlation and Uncertainty

Martin W. McCann, Jr.[1]

[1]Jack R. Benjamin & Associates, Inc., Menlo Park, CA; Stanford University, Stanford, CA

This paper looks at issues associated with estimating risk of seismically initiated failure of dams and other critical infrastructure facilities that are co-located with respect to each other and the location of future earthquakes. In the case of dams, a seismically initiated failure produces a second hazard, the uncontrolled release of the reservoir, which may impact other dams downstream or other critical facilities located on the waterway. Dams may be on the same river or regionally co-located on different rivers. Failure of one or more dams could lead to serial dam failures on a single river; or produce regional consequences on multiple rivers. In the case of downstream critical infrastructure that may be located in the floodway and also impacted by the same seismic event, the ability to prepare for and respond to potential flood conditions may be compromised by the earthquake. This paper looks at the effect of different sources of correlation or dependence on the risk to a portfolio of co-located dams and other critical infrastructure projects. A parametric study is performed to examine the effect that different sources of uncertainty and correlation have on seismic risk analysis results. Sources of uncertainty and correlation that are discussed include the epistemic uncertainty in the seismic hazard analysis, the inter-event earthquake variability and the intra-event spatial correlation of ground motions in the region impacted by the earthquake. The parametric study that evaluates a number of these factors is described. The consequences of a seismic event are not limited to dam failure and downstream inundation. Since it is far more likely that dams may be damaged (as opposed to failing immediately) as a result of the seismic event, the post-event period may be critical in terms of surviving/avoiding uncontrolled release. These 'other' consequences include loss of function (i.e., hydropower production, control of gate systems or other outlets, etc.) as reservoirs may need to be lowered following the earthquake in order to prevent a subsequent failure (e.g., 1971 Lower San Fernando Dam), costs of replacement power or water, unavailability of other infrastructure (i.e., roadways over the dam, etc.).

References

Boore D.M., Gibbs J.F., Joyner W.B., Tinsley J.C., and Ponti D.J. 2003, "Estimated Ground Motion from the 1994 Northridge, California, Earthquake at the Site of the Interstate 10 and La Cienega Boulevard Bridge Collapse, West Los Angeles, California," *Bulletin of the Seismological Society of America* 93(6):2737–2751, http://daveboore.com/pubs_online/bssa_i10.pdf.

Chiou, B.S.-J. and R.R. Youngs, 2008, "An NGA model for the average horizontal component of peak ground motion and response spectra," *Earthquake Spectra* 24(1):173–215, http://dx.doi.org/10.1193/1.2894832

Federal Emergency Management Agency (FEMA), 2009, "HAZUS®MH MR4 Technical Manuals" for the "Earthquake Model", "Hurricane Model," and "Flood Model," Washington, DC, http://www.fema.gov/library/viewRecord.do?id=3731, http://www.fema.gov/library/viewRecord.do?id=3729, and http://www.fema.gov/library/viewRecord.do?id=3726 respectively.

Federal Energy Regulatory Commission (FERC), 2009, "The Strategic Plan: FY 2008–FY 2014," Washington, DC.

Park, J., J.P. Bazzurro, and J.W. Baker, 2007, "Modeling spatial correlation of ground motion Intensity Measures for regional seismic hazard and portfolio loss estimation," *Proceedings of the 10th International Conference on Applications of Statistics and Probability in Civil Engineering*, J. Kanda, T. Takada, and H. Furuta (eds,), Taylor & Francis Group, London, UK, http://www.stanford.edu/~bakerjw/Publications/Park%20et%20al%20(2007)%20Spatial%20Corr elation,%20ICASP10.pdf.

Reed, J.W., McCann, Jr., M.W., Iihara, J. & Hadid-Tamjed, H. 1985. Analytical techniques for performing probabilistic seismic risk assessment of nuclear power plants," *4th International Conference on Structural Safety and Reliability*, No. III, 253–263.

U.S. Nuclear Regulatory Commission, 1997, "Recommendations for Probabilistic Seismic Hazard Analysis: Guidance on Uncertainty and Use of Experts," NUREG/CR-6372, Vol. 1, "Main Report," and Vol. 2, "Appendices," Agencywide Documents Access and Management System (ADAMS) Accession Nos. ML080090003 and ML080090003 respectively.

9.5.4 Storm Surge - Riverine Combined Flood Events

Joost Beckers[1] and Ferdinand Diermanse[1]

[1]Deltares, Inland Water Systems, Delft, the Netherlands

River deltas are flat areas between riverine and coastal influences. These deltas are at risk of flooding caused by intense rainfall runoff, by a severe storm surge at sea, or by a combination of the two phenomena. This implies that there is no single design storm, but rather a collection of likely and less likely combinations of extreme and less extreme rainstorms and storm surges. Together, these combinations constitute the design flood levels. To calculate these flood levels requires a probabilistic assessment of all possible hydraulic conditions and associated probabilities of occurrence. This is a complex and computationally demanding task, especially if the number of combinations is further increased by (for example) the possibility of the failure or nonfailure of a storm surge barrier. In the probabilistic flood hazard assessment, typically hundreds but up to thousands of combinations of external forcings are considered. The design water level at a location of interest is computed for each combination by a hydraulic model. A numerical integration over all possible combinations and their probabilities produces the design flood level for a given return period. The probability of occurrence of a particular combination should take into account the correlation—if any—between the external forcings. Various correlation models can be employed to determine the likelihood of each combination of forcing conditions. An example will be presented in which a copula function was used.

References

Beckers, J.V.L., F.L.M. Diermanse, A. Verwey, et al. ., 2012, "Design of Flood Protection in Hong Kong," *Proceedings of the 2nd European Conference on Flood Risk Management (FLOODrisk 2012)*, Rotterdam, the Netherlands.

Diermanse, F.L.M., and C.P.M. Geerse, 2012, "Correlation models in flood risk analysis," *Reliability Engineering and System Safety* 105:64–72, http://dx.doi.org/10.1016/j.ress.2011.12.004

9.5.5 Combining Flood Risks from Snowmelt, Rain and Ice – The Platte River in Nebraska

Douglas J. Clemetson, P.E.[1]

[1]USACE Omaha District, Omaha, NE

In the past it has been common practice to combine open water and ice-affected flood risks to obtain an all-season flood risk. The all-season stage frequency curve is normally computed using the "probability of a union" equation by adding the probabilities of the open water stage frequency curve and the ice-affected stage frequency curve and subtracting the joint probability of flooding occurring from both types of event. This approach works fine for streams in which both types of events occur every year. However, for the Platte River in Nebraska, ice-affected stages do not occur every year so using this approach would tend to overestimate the frequency of flooding. Therefore, an alternative approach was developed and was subsequently adopted by FEMA for use in Flood Insurance Studies.

The alternate approach involves developing a stage-frequency curve using all annual-maximum stages that are ice-affected events and a separate stage-frequency curve using all the annual-maximum stages that are open water events during the snowmelt and ice season. Each frequency curve is called a "conditional-frequency curve." The ice-affected conditional-frequency curve is "conditioned" in the sense that only annual-maximum peak stages that are related to ice effects are used in the frequency analysis. To obtain the probability of an ice-affected event exceeding a given stage in any year, the exceedance probabilities from the conditional-frequency curve are multiplied by the fraction of time that ice-affected events produce annual-maximum peak stages during the snowmelt/ice season. The open water conditional-frequency curve is "conditioned" in the sense that only annual-maximum peak stages that are open water events are used in the frequency analysis. To obtain the probability of an open water event exceeding a given stage in any year, the exceedance probabilities from the conditional-frequency are multiplied by the fraction of time that open water events produce annual-maximum peak stages during the snowmelt/ice season. The conditional frequency curves for the snowmelt/ice season frequency curve are then combined to obtain the probability of the annual-maximum stage exceeding a given stage in any year as a result of either open water or an ice-affected event during the snowmelt/ice season. For the annual-maximum series, the stage-frequency curves for snowmelt/ice season and the rainfall season are then combined to obtain the all-season frequency curve.

References

U.S. Army Corps of Engineers, "Hydrologic Frequency Analysis", Engineer Manual 1110-2-1415, Washington, DC, March 1993, http://140.194.76.129/publications/eng-manuals/EM_1110-2-1415_sec/EM_1110-2-1415.pdf.

U.S. Army Corps of Engineers, "Hydrologic Analysis, Lower Platte River, Nebraska, Flood Insurance Study," Technical Support Data, Omaha, NB, March 1998.

Federal Emergency Management Agency, "Guidelines and Specifications for Flood Hazard Mapping Partners, Appendix F: Guidelines for Ice-Jam Analysis and Mapping," Washington, DC, February 2002, http://www.fema.gov/library/viewRecord.do?id=2206.

9.5.6 Human, Organizational and Other Factors Contributing to Dam Failures

Patrick J. Regan[1]

[1]Principal Engineer, Risk-Informed Decision Making, Federal Energy Regulatory Commission

"We pretend that technology, our technology, is something of a life force, a will, and a thrust of its own, on which we can blame all, with which we can explain all, and in the end by means of which we can excuse ourselves."
(T. Cuyler Young, *Man in Nature*)

A review of investigation reports for several recent failures of dams, pipelines, and offshore oil platforms suggests that there are many human and organizational factors that contribute to these failures in addition to the, often more simplistic, "root cause" that generally focuses on a single path of component failures and human errors. Among the contributing factors in infrastructure failures are:

1. Organizations being driven by financial performance resulting in decisions being made to operate systems near the boundaries of safety

2. Organization safety programs focused on personal safety rather than system safety

3. A tendency to fix symptoms of problems rather than determine the underlying causes and fix the fundamental problems

4. Organizations that are often complacent about safety because of the low frequency of adverse events, arrogant about the probability of an adverse event happening to them, and ignorant of the real risks inherent in their operations

5. Poor communication within, and outside of, organizations

6. Organizations focused on meeting the letter of regulatory requirements but ignoring the underlying spirit and purpose of the regulations

7. A lack of corporate safety culture in many organizations

The criticality of including factors such as those listed above derives from the fact that the way our infrastructure systems are built and operated today is significantly different from the way similar structures were constructed and operated in the not so distant past. Nancy Leveson, in *Engineering a Safer World*, outlines nine reasons why we need new ways to assess the safety of our systems:

1. Fast pace of technology change
2. Reduced ability to learn from experience
3. Changing nature of accidents
4. New types of hazards
5. Increasing complexity and coupling
6. Decreasing tolerance for single accidents
7. Difficulty in selecting priorities and making tradeoffs

8. More complex relationships between humans and automation
9. Changing regulatory and public views of safety

Leveson also provides considerations for a new accident model to help us understand the complex systems that comprise our infrastructure:

1. Expand accident analysis by forcing consideration of factors other than component failures and human errors.

2. Provide a more scientific way to model accidents that produces a better and less subjective understanding of why the accident occurred and how to prevent future ones.

3. Include system design errors and dysfunctional system interactions.

4. Allow for and encourage new types of hazard analyses and risk assessments that go beyond component failures and can deal with the complex role software and humans area assuming in high-tech systems.

5. Shift the emphasis in the role of humans in accidents from errors (deviations from normative behavior) to focus on the mechanisms and factors that shape human behavior (i.e., the performance-shaping mechanisms and context in which human actions take place and decisions are made).

6. Encourage a shift in the emphasis in accident analysis from "cause" – which has a limiting, blame orientation – to understanding accidents in terms of reasons (i.e., why the events and errors occurred).

7. Examine the processes involved in accidents and not simple events and conditions.

8. Allow for and encourage multiple viewpoints and multiple interpretations when appropriate.

9. Assist in defining operational metrics and analyzing performance data.

In summary, we must understand the broader system in which our infrastructure operates. That system necessarily must include the organizations and people who own operate and maintain these facilities. The purpose of this workshop is to explore these concepts. In many cases the ability of a nuclear station to withstand a flood involves a complex interaction between two very complex systems, each of which is also affected by other systems such as the electric grid, the road network, communication lines, etc.

PANEL 9

SUMMARY OF SIGNIFICANT OBSERVATIONS, INSIGHTS AND IDENTIFIED OPPORTUNITIES FOR COLLABORATION ON PFHA

Co-Chairs:
Thomas Nicholson, NRC
Christopher Cook, NRC

Technical Reporters:
Wendy Reed, NRC

10. Summary of Significant Observations, Insights, and Identified Opportunities for Collaboration on PFHA

Thomas Nicholson[1], Wendy Reed[1], and Christopher Cook[2]

[1]Office of Nuclear Regulatory Research, U.S. NRC, Rockville, MD
[2]Office of New Reactors, U.S. NRC, Rockville, MD

The purpose of the Panel 9 session was to provide an opportunity for the co-chairs and rapporteurs from the eight technical sessions to summarize their session's observations and insights. In particular, the panel co-chairs made short presentations that highlighted the more significant points identified during the panel presentations and discussion.

10.1 Panel 1: Federal Agencies' Interests and Needs in PFHA

Commissioner George Apostolakis provided the keynote address which set the stage for the rest of the workshop. His presentation outlined the importance of risk-informed, performance-based regulations. The Commissioner highlighted the important interplay between probabilistic assessments and traditional deterministic methods; for example, the application of the defense-in-depth concept and safety margins. Management of uncertainties has always been a concern. A risk-informed approach uses a combination of traditional and risk-based approaches through a deliberative process. This process needs to systematically address new data and insights from experience and analyses, including lessons learned from past events and lessons learned from other agencies and the broader technical community. Uncertainties can be identified and estimated using a risk-informed approach.

Panel 1 focused on the participating Federal agencies' interests and needs regarding probabilistic flood hazard assessment (PFHA). Presentations from the Federal agencies highlighted their efforts in developing, using, and implementing probabilistic methods for flood assessments. The first presentation by Fernando Ferrante, of the NRC, described the U.S. Nuclear Commission's (NRC's) probabilistic risk assessment methods and needs in PFHA. It was highlighted that, for nuclear facilities, there can be many different initiating events. How you treat each one may have an effect on plant operations or even other initiating events. These considerations when performing risk analysis may be unique to the NRC. The second presentation by Annie Kammerer, NRC, provided an overview of a comparable approach for estimating seismic event frequencies and effects using an expert elicitation process. The NRC established the Senior Seismic Hazard Analysis Committee (SSHAC). The SSHAC process provides an application of a Probabilistic Seismic Hazard Assessment to another natural hazard, and its relationship to tsunami assessments. The panel discussion mentioned the transferability of the SSHAC process for PFHA.

Like the NRC, the Bureau of Reclamation's (BoR's) dam safety practices involve a mixture of risk-informed and deterministic approaches. BoR conducts risk analysis at different levels, from screening analysis by an individual with peer review input for a comprehensive facility review, up to a risk analysis team which incorporates field personnel. BoR's risk analysis for flood hazard assessments is evolving. The Federal Energy Regulatory Commission (FERC), on the other hand, is currently using a deterministic approach, including establishing a probable maximum flood, but is trying to incorporate risk into its regulatory framework. The industry standards development work by the American Nuclear Society (ANS) is important because it focuses on development of guidance and its implementation through peer-reviewed codes and standards.

<u>Panel 1's Key Observations</u>

- Risk-informed approaches are being used by several Federal agencies and incorporated in safety assessments.

- Risk and deterministic approaches are complementary.

- The Senior Seismic Hazard Analysis Committee (SSHAC) approach is viable; using a comparable expert elicitation process to inform the PFHA process may be a good idea to fill in information gaps such as formulation of flood event scenarios and estimation of their frequencies.

- There are technical challenges with PFHA, but these are being met by the Federal agencies.

- Impediments to implementing PFHA are a willingness to try, the lack of experts, and communication challenges (e.g., use of annual exceedance probabilities in place of N-year flood terminology).

- Need for multidisciplinary teams (e.g., hydrologists, hydrometeorologists, risk analysts, and geomorphologists).

- There is a need for greater incorporation of risk analysis into the education system; however, industrial engineering departments have the expertise in statistics and systems approaches which can be adapted for different environments in which risk assessments are needed.

- There is a need for collaboration among the Federal agencies on PFHA.

<u>Path Forward</u>

- Establish differences and commonalities in risk-informed Federal agency approaches.

- Collaborate to identify potential future activities; involve other stakeholders in the process.

- Consider development and implementation, in the short term, of an SSHAC-type expert elicitation process to fill technological gaps.

10.2 Panel 2: State of the Practice in Identifying and Quantifying Extreme Flood Hazards

Panel 2 focused on the state of the practice in identifying and quantifying extreme flood hazards, including their frequency and associated flood conditions within a risk context. Traditionally, extreme events have been thought of as those occurring well beyond the 500-year flood (0.2 percent annual exceedance probability (AEP) per year). For purposes of this discussion, extreme events have an AEP of 10^{-2} to 10^{-3} per year with "extremely" extreme events having an AEP of 10^{-4} to 10^{-6} per year. It is important to consider the full range of floods out to the extremely rare events, and even the so-called "black swan" events (extremely rare events whose existence is sometimes only acknowledged after they occur).

The major topic of this session was the question of uncertainties. Aleatory uncertainties (i.e., randomness-related uncertainty which cannot be reduced by more sampling) were highlighted in particular as a limit to our forecast ability. For rare events, it was demonstrated that epistemic uncertainties (i.e., structural or conceptual model-related uncertainty which can be reduced by more sampling and analysis such as Bayesian analysis) are more important, and need to be considered. It is thought that Bayesian modeling may be necessary for more extreme events, because this will facilitate determination of model uncertainties. There is also a need to determine the rarity and complexity of natural events, including various combined-event scenarios which may adversely affect structures, systems, and components (e.g., nuclear power plants and dams).

It was emphasized that current approaches to quantifying extreme flood hazards have worked well. The determinations have depended on the site, rarity of the event, and the complexities have varied immensely. Great stochastic approaches already exist; for example, that used to underlie the national flood insurance program. One of the issues with current approaches is that too much focus may be placed on something that is not actually important, such as peak discharge or volume. For example, levees can get wet over time and fail because of that reason, not because of the peak.

The usefulness of paleoflood data was also emphasized. The data is invaluable for providing information regarding historical floods and extending our knowledge to a greater timeframe. In comparison to stream gauge data that can cost approximately $15,000 for one year of data, the price per year of paleoflood information is much less.

Questions that were raised were:

- How do we account for climate change?
- How do we compute uncertainties correctly and with precision?
- How do we ensure consistency in terminology?

10.3   Panel 3: Extreme Precipitation Events

Panel 3 focused on extreme precipitation events and their impacts on flooding caused by local or watershed-shed responses. Many questions regarding extreme precipitation exist:

- How far can one extrapolate data with regards to time intervals?
- What data sets should be used for a particular site?
- What technique to use?

It was also recognized that although the Probable Maximum Precipitation (PMP) is just one point on a curve, such design-basis events can provide useful information for assessing the potential of flooding. It was identified that opportunities exist to collect additional point data to fully characterize the time series of the precipitation distribution. The workshop attendees were informed that NWS Technical Paper 29, which provides point rainfall data, has been superseded and updated by NOAA Atlas 14 (NOAA/NWS, 2004). Recent opportunities discussed were the use of radar data which provide better spatial and temporal correlations of rainfall where rain gauges are not available. It was recognized that much of the foundation of what was presented in the panel session had its basis in reports published by EPRI in the 1993–94 timeframe (EPRI, 1993). These reports contained regional precipitation frequency in the Michigan and Wisconsin areas.

Adequately accounting for uncertainties was identified as a challenge, as was mapping larger regions. Orographic uplift and spatial resolution for radar imaging was identified as a major challenge. There is a pressing need to develop and document an extreme storm catalogue. There has been significant progress in physical and numerical modeling. Models with better resolution combined with more modern data sets, higher-density rain gauges, satellite data, and aircraft observations have all contributed to improved modeling. Information and data from the National Lightning Detection Network (NLDN, http://gcmd.nasa.gov/records/GCMD_NLDN.html) have also been used. Models are also being used for hypothesis testing.

Key barriers that exist are both financial and technical. All Federal agencies are struggling to obtain updated PMP estimates because of funding constraints. NWS is not updating their Hydrometeorological Reports (HMR, http://www.nws.noaa.gov/oh/hdsc/studies/pmp.html). Rather, NOAA is working on their Precipitation-Frequency Atlas 14 which does not yet cover all areas of the United States. Computing times can be long, running into several days for some simulations, though this is something that has been contended with for many years. One way to overcome this is to be more conservative in the use of complex simulation methods; use simplified methods and replace with more complex ones when necessary.

The key observations, insights and opportunities were:
1. Opportunities in Extreme Rainfall Observations/Databases:
   - Continue to obtain and enhance point rainfall data
   - Continue to use the growing availability of radar data, for better spatial and temporal resolution and correlations with point data
   - Need to develop a national extreme storm catalog.

2. Advances in statistics and data processing methods:
   - Regionalization techniques
   - Storm spatial and temporal patterns
   - Mapping larger regions, accounting for seasonal variability
   - Uncertainty estimates.

3. Physical and Numerical Modeling:
   - Use radar and better resolution models to provide better results.
   - Use models for hypothesis testing.
   - Evaluate past events (September 1970 and May 2010 Nashville, TN, storms).
   - Recognize that lack of funding restricts research.

4. Technical and Other Barriers:
   - Technical complexities (e.g., watershed size, different storm mechanisms)
   - Computing resources
   - Skilled personnel
   - Funding

Questions addressed during Panel 3 discussions:

1. Are there opportunities for advances and improvements in extreme rainfall and precipitation observations and databases over the past 30 years to be used for extreme precipitation analyses and their applications for critical infrastructure?

   - Point rainfall data
   - 30 additional years of data are useful

- Radar data: capabilities and limitations
- Better spatial and temporal correlations
- Consider whether intense events underestimated
- Consider whether the useful range of data overestimated

2. How might advances in statistical and data-processing methods be used for extreme precipitation frequency estimates, including evaluating uncertainty?

- Fitting generalized extreme value (GEV) distribution to bound shape parameters
- Changes in regional analyses
- Mapping a large heterogeneous region into multiple homogeneous regions
- Account for seasonal variability
- Use confidence intervals (CI) to indicate statistical level of confidence
- Consider whether CI can be used to address uncertainties

3. How can advances in physical and numerical modeling be applied to provide practical limits to extreme storm scaling?

- Radar and better resolution models provide better results
- Storm transposition
- Test hypotheses
- Evaluation of historic events, such as the May 2010 Nashville, TN, flood (Moore and others, 2012)
- Shortage of funding restricts research, improvement of models, and advances in techniques

4. Are there technical barriers to fully implementing probabilistic extreme storm estimation?

- Technical complexities
- Watershed size and heterogeneity
- Combining the effects of different types of storms
- Combining storm rainfall with snow melt
- Access to computing resources
- Availability of skilled personnel
- Funding

Observations

1. The National Research Council (NAS, 1994) report on extreme precipitation suggested research in several areas, including: radar hydrometeorology and storm catalog, numerical modeling of extreme storms in mountainous regions, and estimating probabilities of extreme storm rainfalls.

2. Numerical models (e.g., WRF) were not intended to analyze extreme rainfall events

3. Results of several case studies have been encouraging

4. Radar data can be used successfully for evaluating extreme events with proper quality control and supplemental rain data

5. How to ensure data independence when using rainfall data (collected in a homogeneous watershed) to define equivalent record lengths

   • understand the watershed
   • avoid oversampling by using a subset of the collected data
   • sort extreme readings above a certain threshold by event date

Next steps that were identified:

1. Focus on extreme rainfall observations and improve databases.

2. Need to explain advances in data-processing methods. The current state of the practice is to use regionalized techniques.

10.4 Panel 4: Flood-Induced Dam and Levee Failures

Panel 4 focused on defining the current state of the art and practice, and research needs, related to estimating probabilities of failure and flooding associated with dams and levees. Observations and insights from the panel session were highlighted. Many of the conclusions overlapped with those from other sessions, such as the need for more reliable data regarding dam and levee failure and increased training of engineers in probability methods and systems engineering for dams and levees.

It was noted that "sunny-day" failures of dams and levees were not addressed during the session. However it was discussed during the audience question and answer period. It was mentioned that dams can also be a source of flood loading. Consideration of this is an important part of the overall assessment process. The Bureau of Reclamation, for example, has tool boxes available which the staff refers to when looking at the "sunny day" failure processes within their risk-assessment process (Wahl, 2004). Many models are available for modeling and estimating dam-breach parameters (Wahl, 2004; Froehlich, 2008; and Xu and Zhang, 2009).

Specific observations and insights are:

1. The probabilistic flood hazard analysis (specifically) and probabilistic risk assessment (PRA) (generally) for dam and levee failure needs a structured evaluation process like the SSHAC process. Because the specific analysis often relies on expert assessments which are biased by human errors, the PRA analysis should evaluate the comprehensive uncertainties of data and modeling.

2. Flood frequency analysis in Bulletin 17B is empirical, being based primarily on observed flood data. But PFHA involves imagining processes by requiring more comprehensive evaluations (as described by SSHAC).

3. Adequate data on large dam failures do not exist. Reliable data on dams, dam components, and operations are generally not available to meet specific needs of risk assessments for individual dams or even components of dam systems.

   a. History shows that large dams seldom fail by flooding alone, but by a combination of factors as a system.
   b. We need more reliable data on dam and component failures and operations.

c. Better data would enable us to reduce subjectivity significantly in risk analysis and decisionmaking.

4. The estimation of dam and levee failures (fragility) is difficult and often subject to bias and limited by engineering tools.

5. The use of risk analyses (conducting comprehensive PFHA) is impeded by the lack of engineers who are trained and are comfortable with the methods that address probability, systems engineering for dams and levees, and risk in high-hazard situations.

6. Probability and risk analysis for dams has come a long way in the past 25 years. Dam owners did not welcome the message of statistics on dam failure in 1975 (general conclusion that $P_{failure} = 1 \times 10^{-4}$).

7. It is generally believed that the chances of dam failure are low (still around 1×10^{-4}, not much change in 40 years), but failures continue to occur. Failures are rarely caused by one factor, but often the "uncommon combination of not uncommon events." This needs to be addressed through risk analysis of dam systems.

8. In some cases, event trees/fault trees cannot handle complexity of events (e.g., examples from Canada and Taum Sauk dam failure events).

9. A lot of problems with data about dam failure events, structure, system and component performance, including the following:

   a. In the past, engineers did not want to talk about failure.
   b. Failure reports tend to only be produced under regulatory requirements.
   c. We do not keep track of near-failure incidents, which are useful to support risk analyses. (In case of failures, we need to look at how we can make the data useful and operational in the future.)
   d. However, some data are available at the following:

      i. RMC (Risk Management Center of USACE) may have some information on reliability of mechanical/electrical components of dams.
      ii. The Centre for Energy Advancement through Technological Innovation (CEATI) Dam Safety Interest Group may be another source of similar information.
      iii. National Performance of Dams Program at Stanford has data on dam incidents.

10. Question was asked if order of magnitude uncertainty in predicting time of failure meant we have a range of 3 min to 30 min or 3 hr. to 30 hr. Reply: It is order of magnitude uncertainty around predicted time of failure from a given regression equation, so it depends on the size of the dam. Point was also made that one big source of uncertainty about failure times is lack of accuracy or consistency in eyewitness reports of dam failure times.

11. Recent experience in laboratory dam breach tests has shown that time of failure (i.e., rate of erosion) for embankment dams is extremely sensitive to erodibility of soils, which is in turn sensitive to compaction processes and conditions during original construction. Tests are now available that can evaluate soil erodibility (in the laboratory or in the field).

12. Historically, dams have done well. The experience worldwide looks good.

13.   Tools have recently become available and continue to be developed that will enable us in the future to add probability and uncertainty to more nodes in the event trees describing sequences of events that lead to dam failure (e.g., process-based erosion and breach models).

14.   Analyses should move toward Bayesian methods. Data are not perfect, but are sufficiently good to have confidence in the analyses performed.

15.   In other fields, society tolerates risks that are not considered acceptable in the dam's arena; potentially develop improved guidance for tolerable risk limits.

16.   The state of PRA in dam and levee safety is relatively new compared to other fields, such as nuclear power plants, where a 2nd or 3rd generation of PRAs is already underway.

17.   Dam failure analysis has been focused heavily on geotechnical engineering practices, but should also incorporate structural, hydrologic, and hydraulic engineering aspects.

18.   A major problem occurs when applying prescriptive methodologies in a "design-safe" manner. This inhibits thinking "outside the box" and hinders advancement in the field. On the other hand, it is costly, risky, and challenging to analyze and tackle challenges in a manner outside the normal convention.

10.5   Panel 5: Tsunami Flooding

Panel 5 focused on tsunami flooding with a focus on Probabilistic Tsunami Hazard Analysis (PTHA) as derived from its counterpart, Probabilistic Seismic Hazard Analysis (PSHA), to determine seismic ground-motion hazards. With regards to establishing a Senior Seismic Hazard Advisory Committee (SSHAC) –type process, which involves expert elicitation, the tsunami community is very transparent, with much being published in the public domain, such as the Princeton Ocean Model (POM) (http://www.aos.princeton.edu/WWWPUBLIC/ htdocs.pom/). There is an understanding within the U.S. community that there is a lack of data on tsunamis. This lack of data stems primarily from the infrequency of tsunamis in the United States compared to storm surge. Unlike Japan, which has many subduction zones, the most likely cause of a tsunami would be from a submarine landslide. The most far field data comes from Hawaii and the maximum height of the data is for a 6- to 9-foot tsunami wave. There are no records for submarine landslides. A lot of imagination is needed to model these occurrences in the United States.

Much more data is needed; however, this can sometimes be hard to obtain because of cost constraints and because much is from foreign countries. Trying to obtain bathymetry data from foreign countries is difficult, for both financial and political reasons. For example, at Tohoku, a boat had to be sent out to measure water displacement. It should also be remembered that an earthquake may not automatically create a tsunami. With regards to the physical processes involved in tsunamis, these are well understood.

Models have been developed that have been peer-reviewed[5].
NRC is a member of an IAEA working group (International Seismic Safety Centre (ISSC-EBP, WA5, WG5-1: Tsunami Hazard))—along with Japan—looking at tsunamis.

Landslide tsunamis, however, require different models which attempt to characterize slip and answer the question of whether or not the land masses break up. A number of institutions are working on this issue, including the University of Rhode Island, University of Delaware, and University of Southern California. Propagation of a tsunami wave up a river is also a concern; however, this is not considered to be an issue as work has been carried out on tidal bores.

The wave effect on structures is not well researched. As previously alluded to, storm surge is the most challenging phenomenon for most of the coastal nuclear power plant sites. What needs to be considered is that even if a plant is designed to withstand a 35-foot storm surge, is there a possibility that, for example, a 16-foot tsunami wave will do more damage because of the different frequency, velocity, or other physical attribute of this wave?

With regards to concerns about tsunamis caused by landslides in the Azores and earthquakes in Iceland and their impacts on the Eastern United States, the NRC has determined that the impacts of such an event on some of the nuclear facilities on the east coast of the United States is not a significant concern[6]. The report was a result of the U.S. Geological Survey being tasked by the Nuclear Regulatory Commission to prepare an evaluation of tsunami sources and their probability to impact the U.S. Atlantic and Gulf of Mexico coasts.

10.6   Panel 6: Riverine Flooding

Panel 6 focused on riverine flooding, including watershed responses via routing of extreme precipitation events and antecedent conditions such as snowpack releases. Mel Schaefer's Stochastic Event Flood Model for Hydrologic Risk Analysis (SEFM) was discussed in detail. Questions included its applicability to larger watersheds (SEFM has been used to model a watershed of 8,000 ft^2 in British Columbia, Canada) and whether it could be used to transpose storm data and account for the areal reduction factor. The point was made that it is not a continuous event model; perhaps a continuous event model would be better at evaluating antecedent moisture and initial conditions in reservoirs. Sequences of floods should also be considered and on this theme of linked flooding events, better communication is needed between reservoir owners when performing watershed modeling with many reservoirs. Mel Schaefer's model (please see http://www.mgsengr.com/Dam_Safety.html) demonstrated reasonable agreement with the PMFs that have been calculated; however, it was suggested that the joint probability method used by the storm surge community may be a better approach for evaluating the uncertainty in the flood discharges. One of the reasons for this is that it

---

[5]   MOST (Method Of Splitting Tsunami), developed originally by researchers at the University of Southern California (Titov and Synolakis, 1998); COMCOT (Cornell Multi-grid Coupled Tsunami Model), developed at Cornell University (Liu and others, 1995); and TSUNAMI2, developed at Tohoko University in Japan (Imamura, 1996). All three models solve the same depth-integrated and 2D horizontal (2DH) nonlinear shallow-water (NSW) equations with different finite-difference algorithms. There are a number of other tsunami models as well, including the finite element model ADCIRC (ADvanced CIRCulation Model for Oceanic, Coastal And Estuarine Waters) (e.g., Myers and Baptista, 1995).

[6]   Atlantic and Gulf of Mexico Tsunami Hazard Assessment Group, "Evaluation of Tsunami Sources with the Potential to Impact the U.S. Atlantic and Gulf Coasts — An Updated Report to the Nuclear Regulatory Commission," U.S. Geological Survey, Administrative Report, August 2008.

enables one to obtain the probability distributions for all of the input parameters and allows one to imagine what the tails look like and to get a wider distribution.

Like other panels, the issues of uncertainties were highlighted. In the case of riverine flooding, how does one address the uncertainties associated with the generation of thousands of years' worth of meteorological data that is used to create peak flow data? What is your uncertainty and how good is it? The point was made that the unit hydrograph method has been used since Sherman developed it in 1932 (Sherman, 1932), so perhaps it is time to move away from this and to use better watershed modeling techniques; for example, the kinematic wave approach (USACE, 2000).

The relation between regulated and unregulated flows can be extrapolated by routing design floods that are extrapolations of historic flows. This approach enables the estimation of exceedance probabilities of extreme floods beyond those observed in the historic record.

10.7   Panel 7: Extreme Storm Surge for Coastal Areas

Panel 7 focused on extreme storm surge for coastal areas caused by hurricanes, extratropical cyclones, and intense winter storms. Joost Beckers' presentation highlighted the 1953 flood in the Netherlands which killed over 1800 people. Many major cities in The Netherlands are below sea level. The Dutch have a very systematic approach to flood research and it was recognized that the United States can learn much from their approach. Stephen Gill described NOAA's recent work and introduced the concept of climate change. Tucker Mahoney described FEMA's coastal hazard analysis; it should be noted that FEMA carries out significant work in this field. Ty Wamsley estimated low-probability events and demonstrated how much coordination is required for a major event. Jen Irish provided an academic perspective on storm surge.

The panel made the following observations:

1. Significant progress has been made in the past 10 years regarding probabilistic analysis in storm surges.

2. Peer review is very important; Federal agencies should make sure to communicate and coordinate with one another.

3. Education in probabilistic techniques is needed. Some schools only teach deterministic concepts in engineering.

4. Federal agencies are cooperating well in the field of storm surge analysis but there is need for improvement.

5. The use of risk-informed decision making can be expanded and information relating to this should be made more readily available.

With regards to issues in this area and future needs, an understanding of upper limits of parameters is required, which can limit the capabilities of flood hazard assessment. For example, uncertainty could lead to probabilistic methods producing a range of central pressures that lead to a 10^{-6} event that are less than was physically thought possible.

Coastal evolution is dynamic and, when combined with sea level rise, requires design and development of coastal facilities that can be constantly adapted. Thought needs to be given as to what can be done to make sure there is resilience to inundation. Shallow-water bathymetry was identified as a source of data that may help with this issue. NOAA currently collects deep-water bathymetry data, but none at lower water levels. Paleo-analysis should also be pursued, even though it will be more difficult to carry out because the coastal environment is constantly changing.

With regards to collaboration amongst agencies:

- There should be greater multi-agency efforts on coordination and guidance between coastal hazard programs. NOAA, NRC, USACE and FEMA have significant programs in this area. 10 years ago, the way in which individual agencies performed coastal hazard assessments was almost independent of each other.

- A database of data and shared codes and modeling capabilities could be created to facilitate the sharing of information between agencies.

10.8 Panel 8: Combined Events Flooding

Panel 8 focused on identifying and evaluating combined event scenarios within a risk-informed framework. The following observations were made:

1. We need to value the work that was done in the past. Many experts that worked on the original ANS standards (though deterministic) will have useful insights.

2. Uncertainties should not be ignored or double-counted, even if they cannot be specifically quantified in the analysis.

3. Not all challenges are technical.

4. It is often the combination of common events, rather than the one extreme rare event, that is the driver for risk.

5. We need to challenge methods and guidance to improve techniques.

6. We should avoid getting precise data with large uncertainties.

7. Focus time and money on things that matter.

10.9 Overall Conclusions

- Risk-informed approaches are being used and are incorporated in safety assessments and decisionmaking by Federal agencies and international groups.

- It is not a question of deterministic versus risk assessment because they are complementary processes. PFHA requires probabilities of initiating events and facilitates uncertainty analysis.

- An expert elicitation strategy similar to the Senior Seismic Hazard Analysis Committee (SSHAC) would help:

    – address paucity of data for characterizing extreme events

- formulate scenarios in hydrometeorologic model simulations
- systematically assess uncertainties (epistemic and aleatory)

The NRC's Commission, in Staff Requirements Memorandum 11-001, requested staff to develop a guidance document to promote the consistent use of expert judgment in regulatory decisionmaking throughout the agency. RES' Human Factors and Reliability branch (HFRB) is leading the effort to develop the requested guidance for staff to use.

- Many of the technical challenges to implement PFHA are being met by researchers; however, impediments to applying PFHA include lack of: data on the frequency and magnitude of the events; willingness to try PFHA; availability of experts; communication; and education.

- PFHA strategies need multidisciplinary teams to:

  - assess complex meteorologic, hydrologic and geologic data
  - simulate hydrologic conditions and scenarios
  - establish and conduct assessments within a risk framework

Future recommended actions to take:

- Advise university and Federal training programs to focus on courses in statistics, risk, and uncertainty assessments.

- Establish understanding of the commonality and differences in risk-informed approaches and decision criteria among the Federal agencies.

- Collaborative and coordinated efforts with other Federal agencies, industry, standards bodies, and other stakeholders to develop mutually accessible data bases and models.

- Consider inclusion of SSHAC-type of approaches for selected hazards to address gaps in data and analytical methods.

- Provide technical staff support to the Advisory Committee on Water Information, Subcommittee on Hydrology work groups related to PFHA issues.

10.10 References

Electric Power Research Institute (EPRI), 1993, "Probable Maximum Precipitation Study for Wisconsin and Michigan: Volume 2: Workbook and User's Guide," TR 101554 V2, Palo Alto, CA.

Froehlich, D. C., 2008, "Embankment Dam Breach Parameters and Their Uncertainties," Journal of Hydraulic Engineering 134(12):1708–1720.

Moore, Benjamin J., Paul J. Neiman, F. Martin Ralph, Faye E. Barthold, 2012, "Physical Processes Associated with Heavy Flooding Rainfall in Nashville, Tennessee, and Vicinity during 1–2 May 2010: The Role of an Atmospheric River and Mesoscale Convective Systems," Monthly Weather Review 140:358–378, http://journals.ametsoc.org/doi/abs/10.1175/MWR-D-11-00126.1.

National Academy of Sciences (NAS), 1994, Estimating Bounds on Extreme Precipitation Events: A Brief Assessment, National Research Council, National Academies Press, Washington, DC.

National Oceanic and Atmospheric Administration/National Weather Service (NOAA/NWS), 2004, NOAA Atlas 14: Precipitation-Frequency Atlas of the United States, Silver Spring, MD, http://www.nws.noaa.gov/oh/hdsc/currentpf.htm.

U.S. Nuclear Regulatory Commission (NRC), 1987, "Methods for the Elicitation and Use of Expert Opinion in Risk Assessment," NUREG/CR 4962.

NRC, 1990, "Eliciting and Analyzing Expert Judgment, A Practical Guide," NUREG/CR 5424.

NRC, 1996, "Branch Technical Position on the Use of Expert Elicitation in the High-Level Radioactive Waste Program," NUREG 1563, Agencywide Documents Access and Management System (ADAMS) Accession No. ML033500190.

NRC, 1997, "Recommendations for Probabilistic Seismic Hazard Analysis: Guidance on Uncertainty and Use of Experts," NUREG/CR 6372, Vol. 1, "Main Report," and Vol. 2, "Appendices," ADAMS Accession Nos. ML080090003 and ML080090004 respectively.

NRC, 2012, "Practical Implementation Guidelines for SSHAC Level 3 and 4 Hazard Studies," NUREG 2117, ADAMS Accession No. ML12118A445.

Olsen, J.R., J. Kiang, and R. Waskom (eds.), 2010, "Workshop on Nonstationarity, Hydrologic Frequency Analysis, and Water Management," Colorado Water Institute Information Series No. 109, Colorado State University, Boulder, CO, http://www.cwi.colostate.edu/nonstationarityworkshop/index.shtml.

Sherman, L.K., 1932, "Streamflow from Rainfall by the Unit Graph Method," Engineering News Record 108:501–505.

Sutley, D.E., Klinger, R.E., Bauer, T.R. and Godaire, J.E. (2009, "Trinity Dam Detailed Hydrologic Hazard Analysis Using the Stochastic Event Flood Model," U.S. Department of Interior, Bureau of Reclamation, Denver, CO, http://odp.trrp.net/FileDatabase/Documents/TrinityDamHydrologicHazFloodModelRpt2.pdf

U.S. Army Corps of Engineers (USACE), 2000, "Hydrologic Modeling System HEC HMS," Technical Reference Manual CPD 74B, Davis, CA, http://www.hec.usace.army.mil/software/hec-hms/documentation/HEC-HMS_Technical%20Reference%20Manual_(CPD-74B).pdf

Wahl, T., 2004, "Uncertainty of Predictions of Embankment Dam Breach Parameters," Journal of Hydraulic Engineering 130(5), American Society of Civil Engineering, pp.: 389–397, http://www.usbr.gov/pmts/hydraulics_lab/pubs/PAP/PAP-0939.pdf.

Xu, Y., and L.M. Zhang, 2009, "Breaching Parameters for Earth and Rockfill Dams," Journal of Geotechnical and Geoenvironmental Engineering 135(12):1957–1970, http://dx.doi.org/10.1061/(ASCE)GT.1943-5606.0000162

APPENDIX A: Workshop Agenda

Agenda

Workshop on
Probabilistic Flood Hazard Assessment
(PFHA)

Time: January 29 – 31, 2013, 8:30 a.m. – 6:30 p.m. (EST)
Location: NRC Headquarters Auditorium, 11555 Rockville Pike, Rockville, MD 20852
WebStreaming: http://video.nrc.gov

Tuesday, January 29, 2013

8:30 a.m.	**Welcome and Introductions** *The Honorable Allison M. Macfarlane*, Chairman, NRC
8:35	**Importance of Risk-Informed, Performance-Based Regulations** NRC Commissioner George Apostolakis
9:00	**Panel 1: Federal Agencies' Interests and Needs in PFHA** Co-Chairs: Nilesh Chokshi, NRC and Mark Blackburn, DOE
9:10	*NRC Staff Needs in PFHA*................................. Fernando Ferrante, NRC
9:30	*Probabilistic Hazard Assessment Approaches: Transferable Methods from Seismic Hazard* .... Annie Kammerer, NRC
9:50	*Reclamation Dam Safety PFHA Perspective*........................John England, BoR
10:10	*FERC's Need for PFHA* ....David Lord, FERC
10:30	Break
10:40	*American Nuclear Society Standards Activities to Incorporate Probabilistic Approaches* ....John D. Stevenson, chair of ANS-2.31; Ray Schneider, Westinghouse
10:55	**Panel 1 Discussion:** Moderators: Nilesh Chokshi, NRC and Mark Blackburn, DOE Rapporteurs: Chris Cook, NRC (NRO) and Marie Pohida, NRC (NRO) Panelists: Fernando Ferrante, Annie Kammerer and Charles Ader, NRC John England, BoR David Lord and Patrick Regan, FERC

Panel Questions

1. What are the roles of deterministic and probabilistic hazard analysis in determining a design basis and conducting a risk assessment? How should they complement each other?
2. What is the status of PFHA? For which flood causing mechanisms PFHAs can be conducted? What improvements are needed for their use in a risk assessment?
3. Given the inherent large uncertainties, how should these be dealt with?
4. What are the impediments, if any, for other flood causing mechanisms to develop PFHA approaches? How they can be overcome?
5. What are your perceptions about the utility and usefulness of a PFHA for your agency missions?
6. Is formal expert interaction approach like SSHAC a viable approach for PFHA? What PFHA specific consideration should be applied?
7. What are the roles of deterministic and probabilistic hazard analysis in determining a design basis and conducting a risk assessment? How should they complement each other?
8. What is the status of PFHA? For which flood causing mechanisms PFHAs can be conducted? What improvements are needed for their use in a risk assessment? How can uncertainties be reduced?
9. What are the impediments, if any, for other flood causing mechanisms to develop PFHA approaches? How they can be overcome?
10. What are your perceptions about the utility and usefulness of a PFHA for your agency missions?
11. Is expert elicitation a viable approach for PFHA?
12. Given the use of PFHA in the development of Design Basis Flooding determination, what is, or should be, the role of Beyond Design Basis Flooding in design and, if required how should it be determined?

11:25   Lunch

12:20 p.m.   **Panel 2: State-of-the-Practice in Identifying and Quantifying Extreme Flood Hazards**
              Co-Chairs: Timothy Cohn, USGS and Will Thomas, Michael Baker, Jr., Inc.

12:25   *Overview and History of Flood Frequency in the United States............................*
              *...*Will Thomas, Michael Baker Corp.

12:50   Keynote: *Extreme Flood Frequency: Concepts, Philosophy, Strategies* ....
              *...*Jery Stedinger, Cornell University

1:15   *Quantitative Paleoflood Hydrology....................................*Jim O'Connor, USGS

1:40   *USACE Methods ....*Douglas Clemetson, USACE

2:05:   *Hydrologic Hazard Methods for Dam Safety ....*John England, BoR

2:30   **Panel 2 Discussion**:
              Moderators: Timothy Cohn, USGS and Will Thomas, Michael Baker, Jr., Inc.
              Rapporteurs: Joseph Giacinto and Mark McBride, NRC (NRO); and

Randy Fedors, NRC (NMSS)
Panelists: Jery Stedinger, Cornell University
John England, BoR
Douglas Clemetson, USACE
Jim O'Connor, USGS

Panel Questions

1. How has the federal agency that you have represented approached the problem of estimating the risk of extreme floods?
2. What is the historical basis for statistical estimation procedures employed by your agency?
3. To what extent are the details of physical processes considered in determining the risk of extreme floods?
4. What criteria are employed in evaluating risk estimation procedures employed by your agency?
5. How could data collection and availability be improved for your agency?
6. What additional data and research are needed to reduce the uncertainty associated with extreme flood frequency estimates?
7. To what extent do operational requirements limit your agency's ability to employ accurate risk estimates?
8. Do fundamentally different issues arise associated with estimating the 50%, 10%, 1%, 0.01%, and 0.0001% exceedance events?
9. Where are the greatest opportunities for improving the way that your agency estimates the risk of extreme floods?

3:00   Break

3:10   **Panel 3: Extreme Precipitation Events**
Co-Chairs: John England, BoR and Chandra Pathak, USACE

3:10   *Introduction of Panel, Objectives, and Questions* ....John England, BoR

3:15   *An Observation-Driven Approach to Rainfall and Flood Frequency Analysis Using High-Resolution Radar Rainfall Fields and Stochastic Storm Transposition.........* ....Daniel Wright, Princeton University

3:40   *Regional Precipitation Frequency Analysis and Extremes including PMP – Practical Considerations*...............Mel Schaefer, MGS Engineering Consultants

4:05   *High-Resolution Numerical Modeling As A Tool to Assess Extreme Precipitation Events*..Kelly Mahoney, NOAA-ESRL

4:30   *Precipitation Frequency Estimates for the Nation and Extremes – A Perspective* .... Geoff Bonnin, NWS-OHD

4:50   *Extreme Precipitation Frequency for Dam Safety and Nuclear Facilities – A Perspective.*...Victoria Sankovich, BoR

5:15   **Panel 3 Discussion**:
Moderators: John England, BoR and Chandra Pathak, USACE

Rapporteurs: Nebiyu Tiruneh, NRC (NRO) and Brad Harvey, NRC (NRO)
Panelists: Daniel Wright, Princeton University
Mel Schaefer, MGS Engineering Consultants
Kelly Mahoney, NOAA-ESRL
Geoff Bonnin, NWS-OHD

Panel Questions

1. Describe the advancements and improvements in extreme storm rainfall and precipitation observations and data bases over the past 30 years. Are there opportunities with radar, point observations, reanalysis data sets, and other data that can readily be utilized for extreme precipitation analyses, understanding, and applications for critical infrastructure?

2. Outline the advances in statistical and data processing methods that can be used for extreme precipitation frequency estimation. These might include regional precipitation frequency, regionalization of parameters, Geographic Information Systems, climatological estimation (such as PRISM), and other areas. How might these tools be applied in practice, and include uncertainty estimates?

3. Describe the advances in physical and numerical modeling of extreme precipitation (such as the Weather Research and Forecasting Model, WRF) that can give insights into the processes and magnitudes of extreme precipitation, including spatial and temporal distributions. How can these tools be applied to provide practical limits for extreme precipitation magnitudes, spatial and temporal storm patterns, transposition, and other extreme storm scaling?

4. The National Research Council (1994) report on extreme precipitation suggested research in several areas, including: radar hydrometeorology and storm catalog, numerical modeling of extreme storms in mountainous regions, and estimating probabilities of extreme storm rainfalls. Are there existing technical barriers to fully probabilistic extreme storm estimation for assessing critical infrastructure, as opposed to Probable Maximum Precipitation?

6:00   Public Attendees' Questions and/or Comments

6:30   Adjourn

Wednesday, January 30, 2013

8:30 a.m.   **Panel 4: Flood-Induced Dam and Levee Failures**
Co-Chairs: Tony Wahl, BoR and Sam Lin, FERC

8:35   *Risk-informed Approach to Flood-induced Dam and Levee Failures.................*
....David Bowles, RAC Engineers & Economists

9:00   *Dutch Approach to Levee Reliability and Flood Risk ....*
.... Timo Schweckendiek, Deltares, Unit Geo-Engineering

9:25   *Risk-Informed Decision-Making (RIDM) Approach for Inflow Design Flood (IDF) Selection and Accommodation for Dams: A Practical Application Case Study.......*
.... Jason Hedien, MWH

9:50	Break

10:45   **Panel 4 Discussion**:

Moderators: Tony Wahl, BoR and Sam Lin, FERC

Rapporteurs: Jacob Philip, NRC (RES); Hosung Ahn, NRC (NRO); and Juan Uribe, NRC (NRR)

Panelists: Eric Gross, FERC, Chicago (Dam)
Timo Schweckendiek, Deltares Unit Geo-engineering (Levee)
Martin W McCann Jr., Stanford University (Dam/Levee)
David Margo, USACE (Dam/Levee)
Gregory Baecher, University of Maryland (Risk Analysis)
Jery Stedinger, Cornell University (Dam)
David Bowles, RAC Engineers & Economists (Dam/Levee)

Panel Questions

1. What does history tell us about the probability of dam/levee failure from various causes?
   - Hydrologic/Hydraulic
     - Overtopping
     - Failure of spillways (concrete lined or earthen/natural rock) or outlet works during extreme floods
     - Seepage and piping
   - Mechanical/Operational
2. What aspects of an event need to have probabilities assigned?
   - Probability of failure itself
   - Probability of flooding of a given severity
3. There are many different triggers for failure. For example, variability of soils and slope protection of embankment dams can be an extreme source of uncertainty. How are the failure thresholds reasonably assumed such as the overtopping depth on an embankment dam based on its existing conditions?
4. What can modeling do to help us understand probabilistic aspects of dam/levee failure?
5. What roles can dam and levee failures play in PFHA or PRA and how can they be integrated? (i.e. make an inventory of relevant issues)
6. Is system behavior (as will be shown in Timo's presentation) relevant for PFHA?
7. What level of detail is appropriate for levee reliability modeling in PFHA?
8. What are the roles of probabilistic safety assessment and probabilistic safety criteria for hydrologic safety of dams?
9. What is different about dam/levee failure impact on downstream as compared to other causes of flooding? For example, dam failure releases large quantities of sediment as well as water, so debris-flow considerations may be more important than for other flooding causes.
10. Do we endorse use of probabilistic methods and risk analysis for the following reasons?
    - Less conservatism than modifying dams for extreme loads
    - Levels playing field and considers relative ranking of all PFMs at a facility
    - Provides a means of prioritizing future studies and modifications
11. Are the issues below potential improvements and research needs or more than that?
    - Spillway debris issues

- Spillway gate failure issues
- River system failures (coordination with other owners)

12. What comes after failure? There are questions, such as: What failure-induced flood parameters or consequences may be important for assessing the safety of assets besides water depth or flow velocities? (e.g., wave impact, scour holes affecting foundations, etc.)

11:35   Lunch

12:35 p.m.   **Panel 5: Tsunami Flooding**
               Co-Chairs: Eric Geist, USGS-Menlo Park, CA and Henry Jones, NRC

12:40   *Probabilistic Tsunami Hazard Analysis* .... Hong Kie Thio, URS Corp.

1:05   *Recent advances in PTHA methodology*...
               ....Randy LeVeque, University of Washington

1:30   *Landslide Tsunami Probability* ....Uri ten Brink, USGS

1:55   *Modeling Generation and Propagation of Landslide Tsunamis*............................
               ....Pat Lynett, USC

2:20   **Panel 5 Discussion**:
               Moderators, Eric Geist, USGS and Henry Jones, NRC
               Rapporteurs: Mark McBride, NRC (NRO) and Randy Fedors, NRC (NMSS)
               Panelists: Randy LeVeque, University of Washington
                           Uri ten Brink, USGS
                           Pat Lynett, USC
                           Annie Kammerer, NRC
                           Frank González, University of Washington
                           Yong Wei, NOAA/PMEL
                           Tom Parsons, USGS
                           Chuck Real, California Geological Survey

Panel Questions

1. What input parameters/uncertainties are important to include in PTHA for extreme tsunamis?

2. What are the appropriate probability distributions (as determined by statistical testing and model selection) that define uncertainty of input parameters for PTHA? What databases exists for earthquakes and landslides to test model distributions and assess uncertainty.

3. What is the best framework (e.g., logic-tree) for including landslide tsunamis in PTHA, given the inherent uncertainties associated with the phenomena? (follow-up to the 2011 NRC/USGS Woods Hole workshop)

4. How does a landslide composed of multiple failures generate a tsunami? By constructive interference of many small failures, only by failure of a dominant large cohesive block, or by hydraulic jump of thick coalesced debris flows from a large area?

5. How do PTHA techniques currently used in the United States differ from those implemented in Japan, especially with regard to extreme tsunami hazards?

6. What are the fundamental needs of improving PTHA (e.g. seafloor mapping/sampling, paleoseismic/paleotsunami deposit analysis, validation of modeled current velocities, etc.), and how should the United States move forward to make these improvements (organizationally and to secure funding)?
7. How should the NRC and other organizations work towards verification and consistency of PTHA methods?

2:50   Break

3:00   **Panel 6: Riverine Flooding**
       Co-Chairs: Will Thomas, Michael Baker, Jr., Inc. and
                     Rajiv Prasad, Pacific Northwest National Laboratory

3:05   *Riverine PFHA for NRC Safety Reviews – Why and How?...*Rajiv Prasad, PNNL

3:30   *Flood Frequency of a Regulated River - the Missouri River...........................*
                 *...*Douglas Clemetson, USACE

3:55   *Extreme Floods and Rainfall-Runoff Modeling with the Stochastic Event Flood*
                 *Model (SEFM)......................................*Mel Schaefer, MGS Engineering

4:20   *Use of Stochastic Event Flood Model and Paleoflood Information to Develop*
                 Probabilistic Flood Hazard Assessment for Altus Dam,Oklahoma....................
                 *... * Nicole Novembre, BoR

4:45   *Paleoflood Studies and their Application to Reclamation Dam Safety.................*
                 *... * Ralph Klinger, BoR

5:10   **Panel 6 Discussion**:
                 Moderators: Will Thomas, Michael Baker, Jr., Inc. and Rajiv Prasad, PNNL
                 Rapporteurs: Peter Chaput, NRC (NRO) and Jeff Mitman, NRC (NRR)
                 Panelists: Douglas Clemetson, USACE
                             Nicole Novembre, BoR
                             Ralph Klinger, BoR
                             Jery Stedinger, Cornell University
                             Mel Schaefer, MGS Engineering

Panel Questions:

1. Runoff simulation-based approaches for riverine PFHA could use either event-based or continuous model simulations. What are the strengths and weaknesses of the two approaches? What R&D is needed to address weakness/gaps?

2. How can we best combine flood frequency analysis approaches (including historical paleoflood information) with simulation approaches to estimate magnitudes and frequencies for extreme flooding events? Is there additional R&D needed in this area?

A-7

3. A full-blown PFHA that includes both sensitivity analysis and uncertainty analysis may be very demanding in terms of computational resources (i.e. large numbers of simulations may be needed). What approaches are available to provide useful results while minimizing the number of simulations that need to be performed?

4. A full-blown PFHA will also be demanding in terms of workload for the analyst. What software tools are available to assist in streamlining the workflow? Is there a significant need for new/improved tools? If so, what is the most critical need?

5. What approaches are available for handling correlations in events/processes that combine to generate extreme riverine floods?

6. In a full-blown PFHA using runoff simulation approach, probability distributions of hydrometeorologic inputs and model parameters are needed. What methods or approaches are available to estimate these probability distributions?

7. Uncertainty in runoff simulations can arise because of uncertainties in inputs, model parameters, and the model structure. What methods or approaches are available to estimate these uncertainties?

8. How do you validate a runoff model for extreme floods?

9. How do you think non-stationarity that has already occurred in the past (e.g., land-use changes) and may occur in the future, (e.g., global climate change) be accounted for in a runoff simulation approach for PFHA?

6:00   Public Attendees' Questions and/or Comments

6:30   Adjourn

Thursday, January 31, 2013

8:30 a.m.   **Panel 7: Extreme Storm Surge for Coastal Areas**
             Co-Chairs: Donald Resio, Univ. of N. Florida and Ty Wamsley, USACE

8:35   *Coastal Flood Hazard in the Netherlands......................Joost Beckers, Deltares*

9:00   *Recent Work and Future Directions in Coastal Surge Modeling within NOAA.......
             .... Stephen Gill, NOAA*

9:25   *FEMA's Coastal Flood Hazard Analyses in the Atlantic Ocean and Gulf of
             Mexico..Tucker Mahoney, FEMA*

9:50   *Modeling System for Applications to Very-Low Probability Events and Flood
             Response...Ty Wamsley, USACE*

10:15   Break

10:30   *Coastal Inundation Risk Assessment* ....
....Jen Irish, Virginia Polytechnic Institute (VPI)

10:55   **Panel 7 Discussion**:
                Moderators: Donald Resio and Ty Wamsley
                Rapporteurs: Mark Fuhrmann, NRC (RES) and Fernndo Ferrante, NRC (NRR)
                Panelists: Joost Beckers, Deltares
                              Stephen Gill, NOAA
                              Tucker Mahoney, FEMA
                              Jen Irish, VPI

Panel Questions

1. Relative to probabilistic hazard prediction, what are some significant advances that your group/organization has made in the last few years or expects to complete in the near future?

2. Relative to probabilistic hazard prediction for NRC, what do you feel could be an important contribution by your group/organization?

3. How should we quantify uncertainty for probabilistic forecasting?

4. What do you feel are some significant shortcomings/lacks of knowledge that could hamper our probabilistic hazard forecasting for the NRC?

5. How can we quantify the effects of upper limits on probabilistic hazards?

11:20   Break

11:25   **Panel 8: Combined Events Flooding**
                Co-Chairs: David Margo, USACE and Joost Beckers, Deltares

11:30   *Combined Events in External Flooding Evaluation for Nuclear Plant Sites*...........
                .... Kit Ng, Bechtel Power Corp.

11:45   *Assessing Levee System Performance Using Existing and Future Risk Analysis Tools* ....Chris Dunn, USACE

12:00   *Seismic Risk of Co-Located Critical Infrastructure Facilities – Effects of Correlation and Uncertainty*..................... Martin McCann, Stanford University

12:15 p.m.   Lunch

1:00   **Panel 8: Combined Events Flooding (continued)**

1:00   *Storm Surge - Riverine Combined Flood Events*.............Joost Beckers, Deltares

1:15   *Combining Flood Risks from Snowmelt, Rain, and Ice – The Platte River in Nebraska* .... Douglas Clemetson, USACE

1:30	*Human, Organizational, and Other Factors Contributing to Dam Failures*............
	.... *Patrick Regan, FERC*

1:45 **Panel 8 Discussion:**
Moderators: David Margo and Joost Beckers, Deltares
Rapporteurs: Michelle Bensi, NRC (NRO) and Jeff Mitman, NRC (NRR)
Panelists: Kit Ng, Bechtel Power Corporation
Chris Dunn, USACE
Martin McCann, Stanford University
Joost Beckers, Deltares
Douglas Clemetson, USACE
Pat Regan, FERC

Panel Questions

1. How can a risk informed framework be utilized to identify plausible event combinations that are relevant to the flood hazard assessment?

2. How can we estimate the probabilities and risks associated with event combinations?

2:30 Break

2:35 **Panel 9: Summary of Significant Observations, Insights, and Identified Opportunities for Collaboration on PFHA**
Panel Co-Chairs: Tom Nicholson and Christopher Cook, NRC
Rapporteurs: Wendy Reed, NRC (RES) and Jacob Philip, NRC (RES)

2:40 Panel 1 Co-Chairs and Rapporteurs

2:55 Panel 2 Co-Chairs and Rapporteurs

3:10 Panel 3 Co-Chairs and Rapporteurs

3:25 Break

3:40 Panel 4 Co-Chairs and Rapporteurs

3:55 Panel 5 Co-Chairs and Rapporteurs

4:10 Panel 6 Co-Chairs and Rapporteurs

4:25 Panel 7 Co-Chairs and Rapporteurs

4:40 Panel 8 Co-Chairs and Rapporteurs

4:55 Workshop and Public Attendees' Questions and/or Comments

6:00 Adjourn

APPENDIX B: Workshop Attendees

The Honorable Allison M. Macfarlane, U.S. Nuclear Regulatory Commission (NRC) Chairman

The Honorable George Apostolakis, NRC Commissioner

Charlie Ader, NRC

Nicholas Agnoli, Federal Energy Regulatory Commission (FERC)

Hosung Ahn, NRC

Scott Airato, FERC

Jon Ake, NRC

Roshan Alayil Divakaran, Atomic Energy Regulatory Board (India)

George Alexander, NRC

William Allerton, FERC

Paul Amico, Hughes Associates, Inc.

Eyal Amitai, NASA GSFC & Chapman University

Richard Anoba, Hughes Associates, Inc.

Rasool Anooshehpoor, NRC

Jeannette Arce, NRC

Hans Arlt, NRC

Taylor Asher, URS Corporation

Thomas Asmus, Nuclear Safety Associates

Mohamad Ali Azarm, IESS/NRC Contractor

Gregory Baecher, University of Maryland

Rakesh Bahadur, SAIC

Daniel Barton, Paul C. Rizzo Associates, Inc.

Edward Beadenkopf, URS Corporation

Scott Beck, Nuclear Safety Associates

Joost Beckers, Deltares

Pradeep Behera, University of the District of Colombia

Joseph Bellini, AMEC/Exelon

Frank Bellini, AREVA

Christopher Bender, Taylor Engineering

Michelle (Shelby) Bensi, NRC

Donald Bentley, Entergy Nuclear Operations

Mihaela Biro, AdSTM

(W.) Mark Blackburn, U.S. Department of Energy (DOE)

Dennis Bley, ACRS

Geoffrey Bonnin, NOAA NWS Office of Hydrologic Development

David Bowles, RAC Engineers & Economists

Sarah Bristol, NuScale Power

Robert Buckley, Savannah River National Laboratory

Shawn Burns, Sandia National Laboratory

Raymond Burski, Florida Power and Light

Christopher Cahill, NRC

Jason Caldwell, Bureau of Reclamation/Technical Service Center

David Capka, FERC

Alice Carson, Bechtel Power

Wei-Wu (William) Chao, TECRO

Li-Chuan Chen, University of Maryland, College Park

Nilesh Chokshi, NRC

Donald Chung, NRC

Jeff Circle, NRC

Doug Clemetson, USACE

Doug Coe, NRC

Mordechai Cohen, NASA Goddard Space Flight Center

Timothy Cohn, USGS

Michael Conway, Constellation Energy Group

Christopher Cook, NRC

Richard Correia, NRC

Brian Cosgrove, NOAA National Weather Service

Lawrence Criscione, NRC

Thomas Cronin, U.S. Geological Survey

Ahmed Dababneh, Paul C. Rizzo Associates, Inc.

Benjamin Daniel, TVA

Paula Davidson, NOAA/National Weather Service

Robert Deppi, FirstEnergy Nuclear Operating Company (FENOC)

Cynthia Dinwiddie, SwRI/CNWRA

Brendan Dooher, PG&E

Raymond Dremel, Maracor Technical Services, Inc.

Claire-Marie Duluc, IRSN

Christopher Dunn, USACE

Angela Duren, USACE

Carville Edwards, FERC

Steven Eide, Scientech/Curtiss-Wright

Michael Eiffe, TVA

John England, BoR

Siamak Esfandiary, DHS/FEMA

Thecla Fabian, IHS McCloskey Nuclear Business

Kenneth Fearon, FERC

Randall Fedors, NRC

Benjamin Ferguson, Paul C. Rizzo Associates, Inc.

Fernando Ferrante, NRC

Allen Fetter, NRC

Cristina Forbes, National Hurricane Center

Robert Fosdick, CNWRA

Stephan Franzone, FPL

Mark Fuhrmann, NRC

Yan Gao, Westinghouse Electric Company

Joseph Gasper, Omaha Public Power District

Eric Geist, USGS

Joseph Giacinto, NRC

Stephen Gill, NOAA/National Ocean Service

Frank González, University of Washington

David Green, NOAA National Weather Service

David Greenwood, Michael Baker Jr., Inc.

Kevin Griebenow, FERC

Allen Gross, NRC

Eric Gross, FERC

Ching Guey, TVA

Jack Guttmann, NRC

Jin-Ping (Jack) Gwo, NRC

James Halgren, Riverside Technology, Inc.

Mark Hammons, FTN Associates, Ltd.

John Hanna, NRC, Region II

Mohammad Haque, NRC

Brad Harvey, NRC

Jason Hedien, MWH Americas, Inc.

Claudia Hoeft, USDA - NRCS

Karin Hollister, Sargent & Lundy, LLC

Victor Hom, NOAA

Donald Hooper, CNWRA

Quazi Hossain, Lawrence Livermore National Laboratory

Michael Houlihan, Geosyntec Consultants

Kenneth Huffman, EPRI

Seung Gyu Hyun, KINS (Korea)

Jennifer Irish, Virginia Tech

Robert Isbell, Duke Energy

Gwo Jack, NRC

Rich Janati, PA Department of Environmental Protection - BRP

Douglas Johnson, FERC

Henry Jones, NRC

David Jula, Michael Baker, Jr., Inc.

Kyle Kaminski, ENERCON

Annie Kammerer, NRC

Lee Kanipe, Duke Energy Corp.

Joseph Kanney, NRC

Julie Kiang, USGS

Dongsoo Kim, NOAA NCDC

Yong Kim, NRC

Yonas Kinfu, Bechtel Power

John Kirkland, NRC

David Kitzmiller, NOAA

Ralph Klinger, Bureau of Reclamation

Paul Knoespel, Nuclear Safety Associates

Gopal Komanduri, Sargent & Lundy LLC

Brenda Kovarik, American Electric Power

Laura Kozak, NRC

Jacob Kwadijk, Deltares

John Lai, NRC/ACRS

John Lane, NRC

Lawrence Lee, Erin Engineering

Michael Lee, NRC

Shizhong Lei, Canadian Nuclear Safety Commission

David Leone, GZA GeoEnvironmental, Inc

Bret Leslie, NRC

Randall LeVeque, University of Washington

Marc Levitan, NIST

Wenwen Li, URS Corporation

Yong Li, NRC

Shyang-Chin (Sam) Lin, FERC

Shielan Liu, BGC Engineering Inc.

Marina Livezey, NOAA/NWS

David Lord, FERC

David Loveless, NRC

Suzanne Loyd, Hughes Associates

Patrick Lynett, University of Southern California

Daniel Mahoney, Private Citizen

Kelly Mahoney, NOAA ESRL/CIRES

Tucker Mahoney, FEMA

Jarrod Malenchak, Manitoba Hydro

Chandrika Manepally, CNWRA

David Margo, USACE

Paul Martinchich, ENERCON

Robert Mason, U.S. Geological Survey

Petr Masopust, AMEC

Peter Mast, Enercon

Thomas Matula, NRC/NMSS

Carl Mazzola, Shaw Environmental Incorporated

Mark McBride, NRC

Martin McCann, Stanford University

Kirsty McConnell, HR Wallingford

Stephen McDuffie, U.S. DOE

Philip Meyer, PNNL

Gerald Meyers, U.S. DOE

Artur Mironenko, Duke Energy

Jeff Mitman, USNRC

Sitakanta Mohanty, CNWRA

Thomas Morgan, Maracor Technical Services, Inc.

Mark Morris, Samui France/HR Wallingford

Lora Mueller, NOAA/NWS

Kit Ng, Bechtel Power Corp.

Thomas Nicholson, NRC

Philippe Nonclercq, EDF

Nicole Novembre, BOR

Jim O'Connor, U.S. Geological Survey

Victor Oancea, SAIC

Ryohji Ohba, NSRA

Ole Olson, Nebraska Public Power District

William Ott, NRC

Thomas Over, USGS

David Passehl, NRC, Region III

Nikhil Patel, Sargent & Lundy LLC

Chandra Pathak, USACE

Malcolm Patterson, NRC

Sanja Perica, NOAA/NWS

Long Phan, NIST/Engineering Laboratory

Jacob Philip, NRC

Joel Piper, Department of Homeland Security

Marie Pohida, NRC

Rajiv Prasad, Pacific Northwest National Laboratory

Qin Qian, Lamar University

John Randall, NRC (Retired)

Harold Ray, ACRS/NRC

Charles (Chuck) Real, California Geological Survey

Vincent Rebour, IRSN

Wendy Reed, NRC

Patrick Regan, FERC

Don Resio, University of North Florida

Jim Riley, NEI

Kurt Roblyer, Enercon Services, Inc.

James Rubenstone, NRC

Gary Ruf, PSEG

William Samuels, SAIC

Selim Sancaktar, NRC

VictoriaSankovich, Bureau of Reclamation/Technical Service Center

John Saxton, NRC

Mel Schaefer, MGS Engineering Consultants, Inc.

Raymond Schneider, Westinghouse

Suzanne Schroer, NRC

Timo Schweckendiek, Deltares

Penny Selman, TVA

Leo Shanley, Erin Engineering and Research, Inc.

Paul Shannon, FERC

Nathan Siu, NRC

Phil Slota, Manitoba Hydro

Eric Smith, Moffatt and Nichol

Heather Smith Sawyer, BWSC

Thomas Spink, TVA

Daniel Stapleton, GZA GeoEnvironmental, Inc.

Jery Stedinger, Cornell University

John Stevenson, J.D. Stevenson, Consulting Engineer

Craig Talbot, Bechtel Power Corporation

Philip Tarpinian, Jr., Exelon Generation

Uri ten Brink, USGS

Hong Kie Thio, URS Corporation

Wilbert (Will) Thomas, Michael Baker, Jr., Inc.

Edward Tomlinson, Applied Water Associates

Richard Turcotte, Nextera Energy-Seabrook Station

Juan Uribe, NRC

Tony Wahl, USBR

Kurt Walter, Mitsubishi

Ty Wamsley, USACE

Hui Wang, University of Texas at Austin

Kenichi Watanabe, NRC

Thomas Weaver, NRC

Michael Weber, NRC

Sunil Weerakkody, NRC

Yong Wei, Pacific Marine Environmental Laboratory

Edwin Welles, Deltares USA

Timothy Wellumson, Xcel Energy

Derek Widmayer, NRC/ACRS

Edmond Wiegert, Mitsubishi Nuclear Energy Systems, Inc.

See Meng Wong, NRC

Daniel Wright, Princeton University

Chung-Sheng Wu, NOAA National Weather Service

Robert Yale, San Onofre Nuclear Generating Station (SONGS)

Kenneth Ying, URS Corporation

Yu Zhang, NOAA

Antonios Zoulis, NRC

APPENDIX C: Acronyms

ACWI	Advisory Committee on Water Information
ADCIRC	ADvanced CIRCulation model (a two-dimensional, depth-integrated, barotropic time-dependent long wave, hydrodynamic circulation model)
AEP	annual exceedance probability
AFP	annual failure probability
ALARA	as low as reasonably achievable
ALARP	as low as is reasonably possible
ALL	annual loss of life
AMR	adaptive mesh resolution
ANCOLD	Australian National Committee on Large Dams
ASCE	American Society of Civil Engineers
BoR	U.S. Department of the Interior, Bureau of Reclamation
Caltrans	California Department of Transportation
CDF	core-damage frequency
CGS	California Geologic Survey
CR	comprehensive review
CSDL	Coast Survey Development Laboratory under NOS/NOAA
D2SI	FERC's Division of Dam Safety and Inspections
DBF	design-basis flood
Deltares	an independent institute in the Netherlands for applied research in the field of water, subsurface and infrastructure with focus on enabling delta life
DOE	U.S. Department of Energy
EEFHA	Expert Elicitation on Flood Hazard Assessment
EMA	Expected Moments Algorithm
EPRI	Electric Power Research Institute

ESEWG	Extreme Storm Event Work Group under ACWI/SOH
ESMF	Earth System Modeling Framework
ESRL	Earth System Research Laboratory under NOAA
ETSS	Extratropical Storm Surge model
FEMA	Federal Emergency Management Agency
FERC	Federal Energy Regulatory Commission
GMPE	ground motion prediction equations
HEC-DSS	Hydrologic Engineering Center Data Storage System
HEC-FIA	Hydrologic Engineering Center Flood Impact Analysis
HEC-RAS	Hydrologic Engineering Center River Analysis System
HEC-SSP	Hydrologic Engineering Center Statistical Software Package
HEC-WAT	Hydrologic Engineering Center Watershed Analysis Tool
HFAWG	Hydrologic Frequency Analysis Work Group under ACWI/SOH
HHA	hydrologic hazard analysis
HAH	high and significant hazard dams
IACWD	Interagency Advisory Committee on Water Data
ICODS	Interagency Committee on Dam Safety
IDF	inflow design flood
IDF	intensity-duration-frequency
INL	Idaho National Laboratory
IPCC	Intergovernmental Panel on Climate Change
IPEEE	Individual Plant Examination for External Events
LGM	last glacial maximum
LIFESim	a model for estimating dam failure life loss
MEOW	maximum envelopes of water

MGB	Multiple Grubbs-Beck test
MHHW	mean higher high water (tide stage)
MLW	mean low water (tide stage)
MSL	mean sea level
NFIP	National Flood Insurance Program
NGO	nongovernmental organization
NLDN	National Lightning Detection Network
NOAA	National Oceanic and Atmospheric Administration
NOAA Atlas 14	Precipitation-Frequency Atlas of the United States
NOS	National Ocean Service of NOAA
NRC	U.S. Nuclear Regulatory Commission
NTHMP	National Tsunami Hazard Mitigation Program
NWS	National Weather Service under NOAA
NWIS	National Water Information System data base
NZSOLD	New Zealand Society on Large Dams
OHD	The Office of Hydrologic Development under NWS/NOAA
PEAKFQ	numerical simulation program for statistical flood-frequency analyses of annual maximum peak flows (annual peaks)
PFDS	precipitation frequency data server
PFHA	probabilistic flood hazard assessment
PFMA	Potential Failure Modes Analysis
PMEL	Pacific Marine Environmental Laboratory
PMF	probable maximum flood (deterministic)
PMP	probable maximum precipitation (deterministic)
PNNL	Pacific Northwest National Laboratory
PRA	probabilistic risk analysis

PPRP	Participatory Peer Review Panel
PSHA	probabilistic seismic hazard assessment
PTHA	probabilistic tsunami hazard assessment
RA	risk assessment
RIDM	risk-informed decisionmaking
ROP	reactor oversight program
SAFRR	USGS Science Application for Risk Reduction
SAPHIRE	Systems Analysis Programs for Hands-on Integrated Reliability Evaluations (an NRC software tool)
SDC	Seismic Design Category
SEFM	Storm Event Flood Model
SLOSH	Sea, Lake, and Overland Surges from Hurricanes model
SOH	Subcommittee on Hydrology under ACWI
SRP	system response probability
SSHAC	Senior Seismic Hazard Analysis Committee
SSSCs	Structures, Systems and Components
SST	sea-surface temperature
SST	stochastic storm transposition
TDIs	technical defensible interpretations
USACE	U.S. Army Corps of Engineers
USGS	U.S. Geological Survey
(U.S.) WRC	Water Resources Council
WRF	Weather Research Forecasting Model

APPENDIX D: Bibliography

Apel, H, A. Thieken, B. Merz and G. Bloschl, "Probabilistic Model for Assessing Flood Risks," *Natural Hazards* 38(1–2):79–100, New York: Springer, May 2006. http://www.springerlink.com/content/64560n3887764083/.

Beckers, J.V.L., Diermanse, F.L.M., Verwey, A., Tse, M.L., Kan, F.Y.F. and Yiu, C.C., "Design of Flood Protection in Hong Kong," Frans Klijn and Timo Schweckendiek, eds., *Flood Risk Management: Science, Policy and Practice: Closing the Gap*, CRC Press, 2013.

Beckers, J.V.L., F.L.M. Diermanse, A. Verwey, et al., "Design of Flood Protection in Hong Kong," *Proceedings of the 2nd European Conference on Flood Risk Management (FLOODrisk 2012)*, Rotterdam, the Netherlands, 2012.

Chen, K. F., "Flood Hazard Recurrence Frequencies for C-, F-, E-, S-, H-, Y-, and Z-Areas," WSRC-TR-99-O0369, Westinghouse Savannah River Company, September 30, 1999, http://www.osti.gov/bridge/servlets/purl/14885-4e3XUm/webviewable/14885.pdf.

Chen, K. F., "Flood Hazard Recurrence Frequencies for A-, K- and L-Areas, and Revised Frequencies for C-, F-, E-, S-, H-, Y- and Z-Areas," WSRC-TR-2000-O0206, Westinghouse Savannah River Company, http://sti.srs.gov/fulltext/tr2000206/tr2000206.html.

Coles, S., L.R. Pericchi, S. Sission, "A Fully Probabilistic Approach to Extreme Rainfall Modelling," *Journal of Hydrology*, 273(1–4):35–50, http://www.sciencedirect.com/science/article/pii/S0022169402003530.

den Heijer, F., and F.L.M. Diemanse, "Towards Risk-Based Assessment of Flood Defences in the Netherlands: an Operational Framework;" F. Klijn and T. Schweckendiek, eds., *Flood Risk Management: Science, Policy and Practice: Closing the Gap*, CRC Press, 2013.

Diermanse, F.L.M., and C.P.M. Geerse, "Correlation Models in Flood Risk Analysis," *Reliability Engineering and System Safety* 105:64–72.

Fontaine, T.A., and K.W. Potter, "Estimating Exceedance Probabilities of Extreme Floods," *Proceedings of the ASCE International Symposium on Engineering Hydrology*, American Society of Civil Engineers, San Francisco, CA, 1993.

Geist, E.L., and U.S. ten Brink, "NRC/USGS Workshop Report: Landslide Tsunami Probability, Convened August 18–19, 2011 at the USGS Woods Hole Science Center, Woods Hole, MA," U.S. Geological Survey, Reston, VA, 2012, http://woodshole.er.usgs.gov/staffpages/utenbrink/my%20publications/WorkshopReport(030112eg).pdf

Global Facility for Disaster Reduction and Recovery, *Risk Analysis Course Manual*, Washington, DC: The International Bank for Reconstruction and Development/The World Bank, 2011, http://www.gfdrr.org/sites/gfdrr.org/files/publication/WB_Risk_Analysis.pdf

Haddad, K., A. Rahman, J.R. Stedinger, "Regional Flood Frequency Analysis Using Bayesian Generalized Least Squares: A Comparison between Quantile and Parameter Regression Techniques," *Hydrological Processes* 26(7):1008–1021, http://onlinelibrary.wiley.com/doi/10.1002/hyp.8189/abstract.

Intergovernmental Panel on Climate Change, "Managing the Risks of Extreme Events and Disasters to Advance Climate Change Adaptation: Special Report of the Intergovernmental Panel on Climate Change," New York: Cambridge University Press, 2012, http://www.ipcc-wg2.gov/SREX/images/uploads/SREX-All_FINAL.pdf

Leander, R., T. A. Buishand, P. Aalders, and M. De Wit, "Estimation of Extreme Floods of the River Meuse using a Stochastic Weather Generator and a Rainfall-Runoff Model," *Hydrological Sciences* 50(6):1089–1103, http://www.knmi.nl/publications/fulltexts/hsj2410_prep.pdf.

Leander, R., and T. Buishand,"A Daily Weather Generator Based on a Two-Stage Resampling Algorithm," *Journal of Hydrology* 374:185–195, http://www.knmi.nl/publications/fulltexts/paper_ngplus.pdf.

Lindenschmidt, K. , U. Herrmann, I. Pech, U. Suhr, H. Apel, and A. Thieken, "Risk assessment and mapping of extreme floods in non-dyked communities along the Elbe and Mulde Rivers," *Advances in Geosciences* 9:15–23, http://www.adv-geosci.net/9/15/2006/adgeo-9-15-2006.pdf

Nott, J., *Extreme Events: A Physical Reconstruction and Risk Assessment*, Cambridge, UK: Cambridge University Press, 2006.

Purvis, M., P.D. Baters, and C.H. Hayes, "A Probabilistic Methodology to Estimate Future Coastal Flood Risk due to Sea Level Rise," *Coastal Engineering* 55(12):1062–1073, http://dx.doi.org/10.1016/j.coastaleng.2008.04.008

Roscoe, K.L., and F. Diermanse, "Effect of Surge Uncertainty on Probabilistically Computed Dune Erosion," *Coastal Engineering* 58(11):1023–1033, http://dx.doi.org/10.1016/j.coastaleng.2011.05.014

Schumann, A.H., ed., *Flood Risk Assessment and Management*, New York: Springer, 2011.

Stedinger, J.R., V.W. Griffis. "Getting from Here to Where?: Flood Frequency Analysis and Climate," *Journal of the American Water Resources Association* 47(3):506–513.

Stedinger, J.R., A.G. Veilleux, and J.R. Lamontagne, "Bayesian WLS/GLS Regression for Regional Skewness Analysis for Regions with Large Cross-Correlations among Flood Flows," *World Environmental and Water Resources Congress 2011: Bearing Knowledge for Sustainability*, American Society of Civil Engineers, Palm Springs, CA, 2011, http://dx.doi.org/10.1061/41173(414)324.

Stedinger, J.R., R.M. Vogel, and E. Foufoula-Georgiou. "Frequency Analysis of Extreme Events," Chapter 18, *Handbook of Hydrology*, D. Maidment (ed.), New York: McGraw-Hill, 1993.

Stedinger, J.R., and D.T. Williams, "Practical Applications of Risk and Uncertainty Theory in Water Resources: Shortcuts Taken and Their Possible Effects," *World Environmental and Water Resources Congress 2011: Bearing Knowledge for Sustainability*, American Society of Civil Engineers, Palm Springs, CA, 2011, http://dx.doi.org/10.1061/41173(414)388.

U.S. Geological Survey, *A Unified Approach to Probabilistic Risk Assessments for Earthquakes, Floods, Landslides, and Volcanoes: Proceedings of a Multidisciplinary Workshop held in Golden, Colorado, November 16–17, 1999*, USGS Open-File Report 01-324, Bismarck, North Dakota, 2001, http://nd.water.usgs.gov/pubs/ofr/ofr01324/index.html.

U.S. Government Accountability Office, "Nuclear Regulatory Commission: Natural Hazard Assessments Could Be More Risk-Informed," GAO-12-465, Washington, DC, April 2012, http://www.gao.gov/assets/600/590431.pdf

U.S. Nuclear Regulatory Commission, "Design-Basis Flood Estimation for Site Characterization at Nuclear Power Plants in the United States of America," NUREG/CR-7046, November 2011, Agencywide Documents Access and Management System (ADAMS) Accession No. ML11321A195.

Wang, Z., and L. Ormsbee, "Comparison Between Probabilistic Seismic Hazard Analysis and Flood Frequency Analysis," *Eos, Transactions, American Geophysical Union* 86(5):45 –47.

APPENDIX E: Electronic Information Sources

Advisory Committee on Water Information (ACWI) Website
http://acwi.gov/

Subcommittee on Hydrology (SOH)
http://acwi.gov/hydrology/index.html

Hydrologic Frequency Analysis Work Group (HFAWG)
http://acwi.gov/hydrology/Frequency/index.html

Extreme Storms Events Work Group
http://acwi.gov/hydrology/extreme-storm/index.html

Federal Emergency Management Agency (FEMA) Website
http://www.fema.gov/

Risk Mapping, Assessment, & Planning
http://www.fema.gov/risk-mapping-assessment-planning

National Flood Insurance Program: Flood Hazard Mapping
http://www.fema.gov/national-flood-insurance-program-flood-hazard-mapping

Coastal Flood Risks: Achieving Resilience Together
http://www.fema.gov/coastal-flood-risks-achieving-resilience-together

Federal Energy Regulatory Commission (FERC) Website
http://www.ferc.gov/

Risk-Informed Decision Making (RIDM)
http://www.ferc.gov/industries/hydropower/safety/initiatives/risk-informed-decision-making.asp#skipnav

Dam Safety and Inspection
http://www.ferc.gov/industries/hydropower/safety.asp

NOAA/National Weather Service (NWS) Website
http://www.weather.gov/

NOAA Atlas 14
http://www.nws.noaa.gov/oh/hdsc/index.html

Precipitation Frequency Data Server (PFDS)
http://dipper.nws.noaa.gov/hdsc/pfds/

NOAA's Climate Program Office
http://cpo.noaa.gov/

NOAA's Climate.gov (Global Climate Dashboard)
http://www.climate.gov/

NOAA's National Geodetic Survey
http://geodesy.noaa.gov/

NOAA's National Oceanographic Data Center (NODC)
http://www.nodc.noaa.gov/

NOAA's National Climatic Data Center (NCDC)
http://www.ncdc.noaa.gov/

NOAA's National Geophysical Data Center (NGDC)
http://www.ngdc.noaa.gov/

NOAA's Satellite and Information Service
http://www.nesdis.noaa.gov/

NWS National Operational Hydrologic Remote Sensing Center
http://www.nohrsc.nws.gov/

NWS Ocean Prediction Center – Extratropical Storm Surge Model
http://www.opc.ncep.noaa.gov/et_surge/et_surge_info.shtml

NOAA's Office of Coast Survey
http://www.nauticalcharts.noaa.gov/

Current NWS Probable Maximum Precipitation (PMP) Documents (HMRs)
http://www.nws.noaa.gov/oh/hdsc/studies/pmp.html

U.S. Drought Portal of the National Integrated Drought Information System (NIDIS)
http://drought.gov/drought/

Great Lakes Environmental Research Laboratory
http://www.glerl.noaa.gov/

U.S. Army Corps of Engineers (USACE) Website
http://www.usace.army.mil/

Flood Risk Management Newsletter
http://operations.usace.army.mil/Flood/pdfs/FRM-1301.pdf

Hydrologic Engineering Center (HEC) for numerical simulation software (e.g., HEC-RAS)
http://www.hec.usace.army.mil/

Advances in Hydrologic Engineering Newsletter
http://www.hec.usace.army.mil/newsletters/HEC_Newsletter_Spring2013.pdf

USACE – Missouri River Basin Water Management Division
http://www.nwd-mr.usace.army.mil/rcc/

USACE – Detroit District – Great Lakes Water Levels
http://www.lre.usace.army.mil/Missions/GreatLakesInformation/GreatLakesWaterLevels.aspx

U.S. Bureau of Reclamation (BoR) Website
http://www.usbr.gov/

Flood Hydrology and Consequences Group
http://www.usbr.gov/pmts/flood/

Best Practices and Risk Methodology
http://www.usbr.gov/ssle/damsafety/Risk/methodology.html

U.S. Department of Energy (DOE) Website
http://energy.gov/

Office of Environmental Management
http://www.em.doe.gov/

DOE Standard "Natural Phenomena Hazards Analysis and Design Criteria for DOE Facilities"
http://www.hss.doe.gov/nuclearsafety/techstds/docs/standard/DOE-STD-1020-2012.pdf

U.S. Geological Survey (USGS) Website
http://www.usgs.gov

USGS Flood Information
http://water.usgs.gov/floods/

USGS Surface Water Information
http://water.usgs.gov/osw/

Water Watch
http://waterwatch.usgs.gov/?m=flood,map&r=us&w=real,map

Probabilistic Analysis of Tsunami Hazards
http://walrus.wr.usgs.gov/reports/reprints/Geist_NH_37.pdf

Nuclear Regulatory Commission (NRC) Website
http://www.nrc.gov/

Workshop on Probabilistic Flood Hazard Assessment
http://www.nrc.gov/public-involve/public-meetings/meeting-archives/research-wkshps.html

NRC's Regulatory Guides
http://www.nrc.gov/reading-rm/doc-collections/reg-guides/

NRC's NUREG-Series Publications
http://www.nrc.gov/reading-rm/doc-collections/nuregs/

NUREG/CR-7046, "Design-Basis Flood Estimation for Site Characterization at Nuclear Power Plants in the United States of America"
http://www.nrc.gov/reading-rm/doc-collections/nuregs/contract/cr7046/cr7046.pdf

NUREG/CR-7134, "The Estimation of Very-Low Probability Hurricane Storm Surges for Design and Licensing of Nuclear Power Plants in Coastal Areas"
http://pbadupws.nrc.gov/docs/ML1231/ML12310A025.pdf

APPENDIX F: Exceedance Probabilities and Associated Recurrence Intervals[1]

Annual Exceedance Probability (AEP)	Recurrence Interval (n-year flood)
0.6667	1.50
0.50	2.00
0.4292	2.33
0.20	5.00
0.10	10.00
0.04	25.00
0.02	50.00
0.01	100.00
0.005	200.00
0.002	500.00
0.001	1,000.00
0.0005	2,000.00
0.0002	5,000.00
0.0001	10,000.00
0.00005	20,000.00
0.00002	50,000.00
0.00001	100,000.00
0.000005	200,000.00
0.000002	500,000.00
0.000001	1,000,000.00

[1] After USGS Fact Sheet FS-2006-3143, Reston, VA, December 2006 (for example, the 500-year recurrence interval flood has an AEP of 0.002)

APPENDIX G: Biographies

George Apostolakis: The Honorable George Apostolakis was sworn in as a Commissioner of the U.S. Nuclear Regulatory Commission (NRC) on April 23, 2010, for a term ending on June 30, 2014. Dr. Apostolakis has had a distinguished career as a professor, an engineer, and risk analyst. He is internationally recognized for his contributions to the science of risk assessment for complex systems. Before joining the NRC, he was a professor of Nuclear Science and Engineering and a professor of Engineering Systems at the Massachusetts Institute of Technology. Dr. Apostolakis served as a member of the NRC statutory Advisory Committee on Reactor Safeguards (ACRS) from 1995 to 2010. He also served as Chairman of the ACRS in 2001 and 2002. In 2007, Dr. Apostolakis was elected to the National Academy of Engineering for "innovations in the theory and practice of probabilistic risk assessment and risk management." He founded the International Conference on Probabilistic Safety Assessment and Management. He served as the Editor-in-Chief of the international journal *Reliability Engineering and System Safety*. Dr. Apostolakis received the American Nuclear Society (ANS) Tommy Thompson Award for his contributions to improvement of reactor safety in 1999 and the ANS Arthur Holly Compton Award in Education in 2005. He is a Fellow of the ANS and the Society for Risk Analysis. Dr. Apostolakis holds a Ph.D. in Engineering Science and Applied Mathematics (awarded in 1973) and an M.S. in Engineering Science (1970), both from the California Institute of Technology. He earned his undergraduate degree in Electrical Engineering from the National Technical University in Athens, Greece, in 1969.

Gregory Baecher: Dr. Gregory B. Baecher is Glenn L. Martin Professor of Engineering in the Department of Civil and Environmental Engineering, University of Maryland. Dr. Baecher was a member of the Interagency Performance Evaluation Task Force (IPET) that performed the risk analysis for post-Katrina New Orleans and is a consultant to the water-resources sector on risk and safety. He is a member of the National Academy of Engineering and the author of four books on risk related to civil infrastructure.

Joost Beckers: Joost Beckers is a senior researcher at Deltares, an independent research institute for water-management issues. Joost holds a Ph.D. in computational physics (1999) and specialized in hydrology and statistics in his professional career. Before joining Deltares in 2005, he was employed at the Dutch Institute for Coastal and Marine Management. At Deltares, he is mainly active in the field of flood risk assessment, with special expertise in extreme value statistics, stochastic modeling, and uncertainty. His recent projects involved loss-of-life risk modeling, establishing design water levels for flood protection, and developing probabilistic models for correlated events.

W. Mark Blackburn: Mark is a twenty-six-year Federal Service employee. Mark is currently Director of the Office of Nuclear Facility Safety Programs (HS-32) within the Department of Energy's (DOE's) Office of Health Safety and Security (HSS) with responsibility for developing and maintaining DOE nuclear facility safety programs in general facility safety, nuclear materials packaging, readiness reviews, oversight of training in nuclear facilities, the DOE facility representative and safety-system oversight programs, and nuclear and nonnuclear facility safety for natural phenomena hazards, fire protection, and maintenance. Before moving into this position, he was Acting Director in the Office of Nuclear Safety Basis and Facility Design with nuclear safety policy responsibility for general nuclear facility safety analysis and design, specific administrative controls, unreviewed safety questions, justifications for continued operations, criticality safety, and nuclear material packaging.

Before joining HSS in July 2009, Mark served in several capacities with the National Nuclear Security Administration (NNSA) from September 1990 through July 2009. As part of the September 2008–May 2009 NNSA Acquisition Strategy Team, he was responsible for identifying and evaluating contracting options for all NNSA sites. As the Pantex Site Office Assistant Manager for Oversight and Assessment and the Chief for Safety, Health, and Quality Assurance, he administered programs in occupational safety and health, construction safety, assessments, operations quality assurance, and weapons and operations quality related to nuclear weapons activities. He also worked as the DOE Albuquerque Operations Office's Nuclear Materials Transportation Manager and as a nuclear weapons manager and engineer at Pantex while gaining a broad management and technical background in operations, safety, quality, and business. Before working at Pantex, he held engineering positions with the Department of Defense (DOD) in Texas, Kentucky, and Missouri.

Mark holds a Bachelor's degree in Industrial Engineering from Mississippi State University and completed the DOD Maintainability Engineering Intern program and NNSA Mid-level Leadership Development Program (MLDP). As part of the MLDP, he briefed the NNSA Administrator on the Pantex Site contractor's year-end performance evaluation and did a rotation at NNSA Headquarters. He is a licensed Professional Engineer in the State of Texas and a Professional Member of the American Society of Safety Engineers.

Geoff Bonnin: Geoff Bonnin is a civil engineer with the National Weather Service (NWS) Office of Hydrologic Development. He manages science and technique development for flood and stream flow forecasting and for water resources services provided by NWS. He initiated and oversees the development of National Oceanic and Atmospheric Administration Atlas 14, the Precipitation Frequency Atlas of the United States, and was lead author of the first three volumes. He was a technical advisor for Engineers Australia's update of their precipitation frequency estimates.

His primary areas of expertise are in the science and practice of real time flood forecasting, estimation of extreme precipitation climatologies (including the potential impact of climate change), data management as the integrating component of end-to-end systems, and the management of hydrologic enterprises. In 2011 he spent several months with the Australian Bureau of Meteorology to identify specific techniques and technologies for sharing between the Bureau and NWS.

David Bowles: David S. Bowles, Ph.D., P.E., P.H., D.WRE, F.ASCE and his colleagues since 1978 have pioneered the development and practical application of risk-informed approaches to dam safety management. They have completed individual and portfolio risk assessments for more than 800 dams in many countries, ranging from screening assessments to detailed assessments with uncertainty analysis. Dr. Bowles has assisted with the development of tailored frameworks for dam and levee safety risk management for government and private owners, regulators and professional bodies in many countries. Clients have included the U.S. Department of the Interior's Bureau of Reclamation ("Reclamation"), U.S. Army Corps of Engineers (USACE), Tennessee Valley Authority, Federal Energy Regulatory Commission, Bureau of Indian Affairs, World Bank, World Meteorological Organization, International Atomic Energy Agency, European Union, Australian National Committee on Large Dams, New South Wales Dams Safety Committee, and numerous private dam owners. He has conducted risk assessments for the failure of dams affecting a nuclear station and currently he advises the Electric Power Research Institute on external flooding probabilistic evaluation and dam failures. Dr. Bowles has served as an expert witness for lawsuits related to dam and canal failures, reservoir operation and hydropower generation, toxic tort, and urban flooding. He has provided training programs on six continents and has authored or reviewed numerous guidance documents for dam safety risk analysis, assessment, and management. He has led software development for dam risk analysis (DAMRAE) and life-loss estimation (LIFESim with a simplified version in HEC-RAS) for USACE, portfolio risk assessment for a large UK dam owner (ResRisk), and realtime reservoir flood operation for Reclamation, USACE, and the Sacramento Area Flood Control Agency. Dr. Bowles is the Managing Principal of RAC Engineers and Economists and an Emeritus Professor of Civil and Environmental Engineering at Utah State University (USU). Previous positions include Director of the Institute for Dam Safety Risk Management and Director of the Utah Center for Water Resources Research at USU, Engineering Department Manager and Branch Manager for Law Engineering's Denver office, and a construction and design engineer for a large international contractor based in the UK.

R. Jason Caldwell: Jason Caldwell is a Meteorologist in the Flood Hydrology and Consequences Group in the Technical Service Center (TSC) of the Bureau of Reclamation in Denver, Colorado. He has served in the TSC for two years and has worked in the fields of meteorology, climatology, and hydrology for over 15 years. His earlier positions were as a Hydrometeorological Analysis and Support Forecaster at the Lower Mississippi River Forecast Center and climatologist at the South Carolina State Climate Office. He will complete a Ph.D. in Civil (Water Resources) Engineering from the University of Colorado in 2013. His principal responsibility is providing expert technical advice on gridded meteorological data processing, hydrometeorological statistics, and extreme storm analysis to TSC management and staff concerning hydrologic issues for Reclamation and other facilities of the Department of the Interior.

Nilesh Chokshi: Currently, Dr. Nilesh Chokshi is Deputy Director of the Division of Site and Environmental Reviews in the Nuclear Regulatory Commission's (NRC's) Office of New Reactors. During his 33 years at the NRC, Dr. Chokshi has managed several research and regulatory areas, including seismic and structural engineering, materials engineering, operating experience risk analysis, and radiation protection. He has also been extensively involved in the area of probabilistic risk assessment, particularly in the development of external event methodology and the standard. Dr. Chokshi has been Vice-Chairman of the Board of Directors of the American Society of Mechanical Engineers' Codes and Standards Technology Institute and a past Chairman of the Committee on the Safety of Nuclear Installations' Working Group on Integrity and Aging of Components and Structures; is currently a member of Advisory Board of the International Association of Structural Mechanics in Reactor Technology (IASMiRT); and was Chairman of the International Scientific Committee of the 16th SMiRT Conference. Before joining the NRC, Dr. Chokshi worked at an architectural/engineering firm involved in designs of nuclear power plants. Dr. Chokshi obtained his Ph. D. in the field of civil engineering (with a specialization in structural engineering) from Rice University and Master's degree from the University of Michigan.

Douglas J. Clemetson: Doug Clemetson serves as the Chief of the Hydrology Section at the Hydrologic Engineering Branch in the Engineering Division of the U.S. Army Corps of Engineers' (USACE's) Omaha District. As Chief of the Hydrology Section, Mr. Clemetson supervises a staff that includes seven hydraulic engineers, a meteorologist, a geographer, and a student, who prepare the hydrologic studies required for the planning, design, and operation of all the water resource projects in the Omaha District.

In addition, his staff advises and supports Emergency Management before and during flood emergencies within the Missouri River Basin. He has extensive experience in flood hydrology, water supply hydrology, statistics, watershed modeling, reservoir simulation, extreme storm analysis, and flood forecasting.

During his career, Mr. Clemetson has worked in the Missouri River Division's Technical Engineering Branch and Reservoir Control Center and has served as an H&H Specialist in the Headquarters USACE Emergency Operations Center following Hurricanes Katrina and Rita in 2005. He has also been the leader of the Inflow Flood Methodology Team for the USACE Dam Safety Program; is the USACE representative on the federal interagency Extreme Storm Events Work Group; is the leader of the USACE Extreme Storm team; and has been a member of the USACE Hydrology Committee since 1996, where he currently serves as vice-chairman of the committee.

Mr. Clemetson earned his bachelor's degree in Civil Engineering from South Dakota State University in 1980. He is a registered Professional Engineer in the State of Nebraska, a member of the American Society of Civil Engineers, and a member of the Association of State Dam Safety Officials. He has been with the Corps of Engineers in Omaha for over 32 years.

Tim Cohn: Tim Cohn works in the U.S. Geological Survey (USGS) Office of Surface Water, where he has co-authored more than 25 papers on methods for estimating flood risk and related topics. He previously served as USGS Science Advisor for Hazards and as the American Geophysical Union's 1995–96 American Association for the Advancement of Science (AAAS) Congressional Science Fellow in the office of Senator Bill Bradley. Dr. Cohn holds M.S. and Ph.D. degrees from Cornell University and a B.A. from Swarthmore College.

Christopher Cook: Starting in December 2012, Dr. Cook became chief of the Hydrology and Meteorology Branch of the Office of New Reactors (NRO) at the Nuclear Regulatory Commission (NRC). Before joining the branch, he also served as chief of the Geoscience and Geotechnical Engineering Branch and a senior hydrologist in the Office of New Reactors for a number of years. As a hydrologist, Dr. Cook performed numerous technical reviews to support NRC's safety evaluation reports (SER) and environmental impact statements (EIS) associated with Early Site Permit (ESP) and Combined License applications for new nuclear reactors. Before joining the NRC, Dr. Cook supported the U.S. Nuclear Regulatory Commission as a technical reviewer in the area of hydrology while employed at the Pacific Northwest National Laboratory (PNNL). He contributed to technical reviews associated with NRC's SERs and EISs on the North Anna, Clinton, Grand Gulf, and Vogtle ESP applications. Past research at PNNL and the University of California–Davis focused on multidimensional hydrodynamic and water quality modeling of surface water systems, including the use of three-dimensional computational fluid dynamics (CFD) models.

Dr. Cook holds a B.S. in Civil Engineering from Colorado State University, a M.S. in Civil Engineering from the University of California–Davis specializing in groundwater hydrology, and a Ph.D. from the University of California–Davis specializing in multidimensional hydrodynamic and water quality modeling of surface water systems.

Christopher N. Dunn: Christopher Dunn is the director of the Hydrologic Engineering Center (HEC) in Davis, CA, with more than 25 years' experience. His expertise includes flood damage and impact analysis, planning analysis, risk analysis, levee certification, river hydraulics, surface water hydrology, storm water management, watershed systems analysis, and ecosystem restoration. He was the project manager for HEC's role in the Sacramento/San Joaquin Comprehensive Study and also led the development of flood damage reduction, ecosystem restoration, and system analysis software tools. He has also worked on several international projects, including water management modeling in Iraq and Afghanistan, training in Japan, and collaboration with the U.S. Geological Survey on a Learning Center in Turkey.

John England: John England is a flood hydrology technical specialist with the Bureau of Reclamation in Denver, CO. Dr. England's research and project interests include extreme flood understanding and prediction, rainfall-runoff modeling, flood frequency, hydrometeorology, paleoflood hydrology, and risk analysis. For the past 15 years he has developed and applied probabilistic flood hazard techniques to evaluate the risk and safety of Bureau of Reclamation dams, and has overseen implementation of risk-based techniques for the dam safety program. Mr. England has authored numerous publications, including journal articles, book chapters, conference proceedings, guidelines, technical manuals, and reports. Dr. England received his M.S. and Ph.D. in hydrology and water resources from Colorado State University. He is a registered Professional Hydrologist with the American Institute of Hydrology, a registered Professional Engineer in Colorado, and holds a Diplomate, Water Resource Engineer (D.WRE), from the American Academy of Water Resources Engineers. Dr. England was named the Bureau of Reclamation's Engineer of the Year and was nominated as one of the top 10 Federal Engineers in 2008.

Fernando Ferrante: Fernando Ferrante is currently working as a Reliability and Risk Analyst with the U.S. Nuclear Regulatory Commission (U.S. NRC) in the application of probabilistic risk assessment (PRA) tools and models for the oversight of operating nuclear reactors in the US. His primary focus is on quantitative risk assessments using the Systems Analysis Programs for Hands-on Integrated Reliability Evaluations (SAPHIRE) software, as well as the development of PRA tools for use in analyzing so-called "external" events (e.g., internal fire, seismic, and external flooding events).

Before joining the U.S. NRC, Dr. Ferrante worked as a Research Engineer at Southwest Research Institute in the Center for Nuclear Waste Regulatory Analysis. His responsibilities included prelicensing activities in support of the U.S. NRC High-Level Waste program; development of risk-informed, performance-based regulatory guidance; and performance assessment of a potential nuclear high-level waste repository.

Dr. Ferrante has a Ph.D. from Johns Hopkins University (2005) in Civil Engineering (thesis topic on probabilistic mechanics), M.S. in Civil Engineering from the University of Virginia (2000), and a B.S. in Mechanical Engineering from University College London, England (1997).

Eric Geist: Eric Geist is a Research Geophysicist with the Pacific Coastal & Marine Science Center of the U.S. Geological Survey, where he has worked for 28 years. Throughout his career, he has focused on computer modeling of geophysical phenomena, including large-scale deformation of the earth in response to tectonic forces and the physics of tsunami generation. For the last ten years, he has led research and developed methods for the probabilistic analysis of tsunami hazards. Mr. Geist has authored over 120 journal articles and abstracts, including an article in *Scientific American* on the devastating 2004 Indian Ocean tsunami and a series of review papers on tsunamis for *Advances in Geophysics*. Eric received his BS degree in Geophysics from the Colorado School of Mines and his MS degree in Geophysics from Stanford University.

Eric Gross: Eric Gross has a Bachelor's degree from Rensselaer Polytechnic Institute and a Master's in Civil/Water Resources Engineering from the University of Maryland. He is registered as a Professional Engineer (Civil) in the State of Maryland. He has been with the Federal Energy Regulatory Commission (FERC) for 10 years and is a member of the FERC's Risk-Informed Decision Making (RIDM) team. The RIDM Team is tasked with incorporating risk concepts into FERC's dam safety program. Currently, he is chairing FERC's Risk Technical Resource Group, heading the team writing FERC's guidelines for dam failure consequences determination, and is a member of the St. Louis River Project canal embankment failure investigation team. He was a member of the Taum Sauk Dam failure investigation team and the lead engineer for FERC's first risk-informed inflow design flood determination.

Jason Hedien: Jason Hedien is Vice President of, and a Principal Geotechnical Engineer and Project Manager with, MWH Americas, Inc. At MWH Jason has over 18 years of experience in the management, inspection, analysis, and design of dams and hydropower projects, infrastructure and waterway/port projects, and wastewater and water supply projects. Mr. Hedien's professional experience includes conducting deterministic and probabilistic seismic hazard analyses, risk assessments, earthquake engineering for dams and civil infrastructure projects, and dam safety engineering for the evaluation of dam safety and implementation of dam safety improvements for dams and ancillary structures.

Jennifer L. Irish: Dr. Irish is an associate professor of civil engineering at Virginia Polytechnic Institute and State University ("Virginia Tech") with expertise in storm surge dynamics, storm morphodynamics, vegetative effects, coastal hazard risk assessment, and general coastal engineering. She is a Diplomate of Coastal Engineering and licensed Professional Engineer with 18 years of experience. Dr. Irish has published more than 30 journal papers. Irish recently received the Department of the Army's Superior Civilian Service Award (2008) and Texas A&M University's Civil Engineering Excellence in Research Award (2010). Dr. Irish teaches coastal engineering and fluid mechanics and leads research on the coastal impacts of extreme events and climate change.

Henry Jones: Henry Jones is a hydrologist in the Division of Site Safety and Environmental Analysis of the Office of New Reactors within the U.S. Nuclear Regulatory Commission (U.S. NRC). He has worked at the U.S. NRC for 6 years in the Branch of Hydrology and Meteorology. Before working at the U.S. NRC, Henry Jones served 28 years in the United States Navy as a physical oceanographer and meteorologist. His principal responsibility is providing expert technical advice to NRC management and staff concerning tsunami and storm surge hydrologic issues for NRC-licensed facilities.

Annie Kammerer: Dr. Annie Kammerer is a senior seismologist and earthquake engineer in the Office of Nuclear Regulatory Research at the United States Nuclear Regulatory Commission (NRC), where she coordinates the NRC's Seismic Research Program. In this role, she is responsible for overseeing research on a broad range of seismic topics ranging from seismic and tsunami hazard assessment to seismic risk assessments for nuclear facilities. She was project manager and contributing author of the NRC's current guidance on conducting seismic hazard assessments (Regulatory Guide 1.208 and NUREG-2117). She is also currently the NRC project manager and representative on the Joint Management Committee for "Next Generation Attenuation-East" project, which is developing ground motion prediction equations for central and eastern North America for use with the recently published Central and Eastern United States Seismic Source Characterization (CEUS SSC) for Nuclear Facilities model. She was a member of the Participatory Peer Review Panel for the CEUS SSC project.
Dr. Kammerer is currently the NRC's technical lead for the seismic walkdowns being conducted at all 104 operating reactors in the U.S. as part of the NRC's post-Fukushima activities.
Dr. Kammerer has authored over three dozen publications, including regulatory guidance, technical reports, journal articles, a book chapter, conference proceedings and papers, and an American Society of Civil Engineers special publication. She was a contributing author on the tsunami guidance contained in the International Atomic Energy Agency's Safety Standard Guide 18.

Dr. Kammerer holds degrees from the University of California–Berkeley, including a B.S. degree in Civil Engineering, a M.S. degree in Geotechnical Engineering, and a Ph.D. in Geotechnical Earthquake Engineering, with minors in Seismology and Structural Engineering.

Ralph Klinger: Ralph Klinger is a Quaternary Geologist and Geomorphologist at the Bureau of Reclamation's Technical Service Center in Denver, Colorado. He received his B.S. and M.S. in Geological Sciences from San Diego State University and his Ph.D. in Geology from the University of Colorado–Boulder. He has been involved in studying the geologic record of natural hazards throughout the western U.S. for almost 30 years. His principal responsibility at Reclamation is to provide technical expertise to the Dam Safety Office and advise the engineering staff on seismic and flood hazards and their use in the evaluation of risk.

Randall LeVeque: Randy LeVeque is a Professor of Applied Mathematics at the University of Washington, where he has been on the faculty since 1985. He is a Fellow of the Society for Industrial and Applied Mathematics and of the American Mathematical Society. His research is focused on the development of numerical algorithms and software for the solution of wave propagation problems. For the past 10 years, he has been working on development and application of the GeoClaw software for tsunami modeling and other hazardous geophysical flows.

S. Samuel Lin: S. Samuel Lin, PhD, P.E., D.WRE is a civil engineer specializing in hydrologic and hydraulic (H&H) safety and associated risk analyses of hydropower generation dams under the Federal Energy Regulatory Commission's jurisdiction. He has been a Federal and State dam safety regulator for 24 years. He also worked in a consulting practice as water resources engineer specializing in planning and H&H analyses for design of dams. He is presently serving on the steering committee for the Federal Emergency Management Agency's update of the national guidelines on Selecting and Accommodating Inflow Design Flood (IDF) for Dams for hydrologic safety of dams. He also serves on the Interagency Committee on Dam Safety's Frequency of Extreme Hydrologic Events Guidance Development Task Group. He was Chair of the federal interagency Subcommittee on Hydrology (SOH) for 2005–2007. For chairing the SOH, he is the recipient of a Leadership Recognition Award and Certificate of Appreciation from the U.S. Geological Survey and Advisory Committee on Water Information, respectively.

David W. Lord: Mr. David Lord, P.E., is a senior civil engineer in the Division of Dam Safety and Inspections (D2SI) in the Portland Regional Office of the Federal Energy Regulatory Commission (FERC), where he has worked since 1985. His first year was in the San Francisco Regional Office. D2SI is responsible for the inspection of dams in the United States regulated by the FERC. The FERC regulates dams that are owned by non-federal operators of hydroelectric projects.

Mr. Lord is currently one of three people responsible for developing the D2SI Risk-Informed Decision Making (RIDM) program. This primarily consists of teaching concepts of risk to FERC staff, owners, and consultants, and coordinating the writing of 25 chapters of new RIDM Engineering Guidelines.

He is a member of the American Society of Civil Engineers, United States Society of Dams, and Association of State Dam Safety Officials. He has authored several papers for dam safety conferences.
Mr. Lord has a M.S. in Civil Engineering from the University of Portland (May 1984) and a B.S. in Chemistry and Biology (double major) from Whitworth College (May 1974).

Tucker B. Mahoney: Tucker Mahoney is a coastal engineer for the Federal Emergency Management Agency (FEMA), Region IV. Ms. Mahoney is the FEMA project manager for coastal Flood Insurance Studies in the southeastern United States. She also participates on coastal policy and coastal outreach forums within FEMA. Before joining FEMA, Ms. Mahoney was a coastal engineer for Moffatt & Nichol, where she completed a variety of hydrodynamic and sediment transport modeling studies and related construction projects.

David Margo: David Margo earned a Bachelor of Science degree in engineering in 1993 and a Master of Science degree in civil engineering in 1995, both from the University of Pittsburgh. He has worked for the U.S Army Corps of Engineers (USACE) for 17 years on various civil works projects. His areas of expertise and professional experience include risk analysis for dams and levees, flood frequency analysis, surface water hydrology, river hydraulics, and levee certification. Mr. Margo has been involved in probabilistic risk studies for a number of dams and levees in the USACE portfolio, including the Dallas Floodway and Bluestone Dam. He is also involved in methodology and policy development for the USACE dam and levee safety programs and is one of the lead developers for the USACE risk-informed levee screening tool.

Martin McCann, Jr.: Dr. McCann received his B.S. in civil engineering from Villanova University in 1975, an M.S. in civil engineering in 1976 from Stanford University, and his Ph.D. in 1980, also from Stanford University.

His areas of expertise and professional experience include probabilistic risk analysis for civil infrastructure facilities and probabilistic hazards analysis, including seismic and hydrologic events, reliability assessment, risk-based decision analysis, systems analysis, and seismic engineering. He currently teaches a class on critical infrastructure risk management in the Civil and Environmental Engineering Department at Stanford.

He has been involved in probabilistic risk studies for critical infrastructure (dams, levees, nuclear power plants, ports, and chemical facilities) since the early 1980s. He has performed probabilistic flood hazard assessments for a number of Department of Energy sites and commercial nuclear power plants. Recently, Dr. McCann led the Delta Risk Management Strategy project that conducted a risk analysis for over 1,100 miles of levee in the Sacramento and San Joaquin Delta. He was also a member of the U.S. Army Corps of Engineers' Interagency Performance Evaluation Task Force (IPET) Risk and Reliability team evaluating the risk associated with the New Orleans levee protection system following Hurricane Katrina.

Dr. McCann developed the SHIP risk analysis software that is used to perform risk and uncertainty calculations for facilities exposed to external hazards.

He is currently serving on the American Nuclear Society (ANS) 2.8 committee that is updating the requirements for the assessment of external flood hazards at nuclear facilities.

Kit Y. Ng: Kit Ng is a senior principal hydraulic and hydrology specialist and also serves as the assistant chief of Bechtel Power Corporation's Geotechnical & Hydraulic Engineering Services Group. She has 22 years of industry experience and is the technical lead in Bechtel Power on hydrologic and hydraulic design and modeling, specifically on flooding hazard evaluation for nuclear power facilities, design of cooling water systems, intake and outfall hydraulic design, water resource management, storm water management, site drainage, erosion and sediment controls, coastal hydrodynamics, and contaminant mixing and transport modeling in both surface water and groundwater. Kit has a Ph.D in environmental hydraulics from the California Institute of Technology. She has served as the chair for the Computational Hydraulic Committee of the American Society of Civil Engineers' Environmental & Water Resources

Institute and is currently a member of the American National Standards Institute and American Nuclear Society (ANSI/ANS) 2.8 and 2.31 standard committees on design basis flood determination for nuclear facilities and estimating extreme precipitation at nuclear facility sites. She is also participating in the new ANS 2.18 standard committee on evaluating radionuclide transport in surface water for nuclear power sites.

Thomas Nicholson: Thomas Nicholson is a Senior Technical Advisor in the Division of Risk Analysis of the Office of Nuclear Regulatory Research within the U.S. Nuclear Regulatory Commission (U.S. NRC). He has served in the research office for 32 years and has worked at the U.S. NRC for 36 years, receiving numerous awards, including the U.S. NRC Meritorious Service Award for Scientific Excellence. His earlier positions were as a senior hydrogeologist and hydrologist/hydraulic engineer in the Offices of Nuclear Regulatory Research, Standards Development, and Nuclear Reactor Regulation.

His principal responsibility is providing expert technical advice to NRC management and staff concerning radionuclide transport in the subsurface at NRC-licensed facilities. As a senior project manager, he has formulated and directed numerous research studies involving estimation of extreme flood probabilities in watersheds; radionuclide transport in fractured rock; and integration of subsurface monitoring and modeling. He presently serves as chair of the Extreme Storm Event Work Group under the Federal Subcommittee on Hydrology of the Advisory Committee on Water Information. He is the NRC liaison to the Water Science and Technology Board of the National Academies of Sciences.

He holds a B.S. in geological sciences (with honors) from Pennsylvania State University and a M.S. in geology (hydrogeology) from Stanford University. At Stanford, he was a student of Professors Ray Linsley and Joseph Franzini in the hydrology program while completing the core courses in hydrology. He is an active member of the American Geophysical Union, American Institute of Hydrology, Geological Society of America, International Association of Hydrological Sciences, the International Hydrogeologic Society, and the National Ground-Water Association, and is a registered Professional Geologist.

Nicole Novembre: Nicole Novembre, P.E., is a hydrologic engineer with the Bureau of Reclamation Technical Service Center's Flood Hydrology and Consequences Group in Denver, Colorado. She has been with Reclamation for 3 years. Previously she spent 2 years in private consulting after earning her Hydrology, Water Resources, and Environmental Fluid Mechanics M.S. degree from the University of Colorado at Boulder in 2007. Her principal responsibility is to develop hydrologic hazard loadings for use in Bureau of Reclamation Dam Safety risk assessments.

Jim O'Connor: Jim O'Connor is a Research Hydrologist at the U.S. Geological Survey's Oregon Water Science Center in Portland, Oregon, USA. He is a U.S. Pacific Northwest native long interested in the processes and events that shape the remarkable and diverse landscapes of the region. Following this interest with a Geological Science major at the University of Washington and M.S. and Ph.D. degrees at the University of Arizona (1990), he has spent the last 23 years focused on floods, fluvial geomorphology, and Quaternary geology, primarily in the western United States.

Chandra Pathak: Dr. Chandra S. Pathak, Ph.D., P.E., D.WRE, F.ASCE, has a distinguished career with over 33 years of experience in wide-ranging areas of water-resources engineering that has included surface and groundwater hydrology and hydraulics; stormwater management; wetland; water quality; drought management; Geographic Information Systems (GIS); and hydrology, hydraulic, and water-quality computer models. Currently, he is a principal engineer at the U.S. Army Corps of Engineers headquarters in Washington, DC. Before that he was a principal engineer at the South Florida Water Management District for 12 years. He is an adjunct professor at Florida Atlantic University and Florida International University. Previously, he was a practicing consulting engineer in the United States for over 20 years. He obtained a bachelor of technology degree in 1976, a master of engineering in water resources degree in 1978, and a doctorate in hydrologic engineering in 1983 from Oklahoma State University. Since 2006, he has been serving as an associate editor of the *Journal of Hydrologic Engineering*. He has numerous presentations, speeches, and technical papers to his credit in the areas of water resources engineering. In 2007, he was awarded Diplomate of Water Resources Engineering and Fellow Member of the American Society of Civil Engineers (ASCE).

Rajiv Prasad: Dr. Prasad has over 20 years of experience as a Civil Engineer. His Ph.D. dissertation at Utah State University focused on spatial variability, snowmelt processes, and scale issues. He has been employed at the Pacific Northwest National Lab (PNNL) for the last 13 years. His research has included application of spatially distributed models for water management, watershed process characterization, and impacts from global climate change. Dr. Prasad has been working on the Nuclear Regulatory Commission's (NRC's) permitting and licensing reviews since the wave of new power reactor applications started around 2003–04. Dr. Prasad has led PNNL's effort in updating the NRC's Standard Review Plan in 2007, developing staff guidance related to tsunamis in 2009, and revising the technical bases for design-basis flood estimation at nuclear power plant sites in 2010–11. Currently, Dr. Prasad is leading the PNNL team tasked with providing technical bases for a comprehensive probabilistic flood hazard assessment at nuclear power plant sites.

Charles R. Real: Charles Real is a Registered Geophysicist in California and has worked in the field of earthquake hazards for over 35 years. He is currently a Supervising Engineering Geologist with the California Geological Survey, where he helped establish and currently manages California's Seismic Hazard Zonation Program. During his career he has been principal investigator for numerous federal grants, and has recently been elected to the Board of Directors for the Northern Chapter of the Earthquake Engineering Research Institute. He currently co-chairs a California Tsunami Policy Working Group, focusing on ways to reduce coastal communities' tsunami risk.

Wendy Reed: Dr. Reed is a radiochemist in the Environmental Transport Branch of the Division of Risk Analysis of the Office of Nuclear Regulatory Research in the U.S. Nuclear Regulatory Commission (U.S. NRC). She has worked at the U.S. NRC for over 4 years. She holds a B.Sc. (Honors) and a Ph.D., both in chemistry, from the University of Manchester in the United Kingdom.

Patrick J. Regan: Pat Regan is the Principal Engineer for Risk-Informed Decision-Making in the Division of Safety of Dams and Inspections, Office of Energy Projects, Federal Energy Regulatory Commission (FERC-D2SI). He has worked in the hydroelectric generation field for 36 years for a dam owner, as a consultant, and (since 2000) for FERC-D2SI. He has authored or co-authored several papers on use of risk assessment in dam safety and has received both the United States Society on Dams Best Paper award and the Association of State Dam Safety Officials' West Region Award of Merit. His principal responsibility is leading the development of the FERC's Risk-Informed Decision-Making program.

Donald T. Resio: Dr. Resio is the Director of the Taylor Engineering Research Institute and Professor of Ocean Engineering within the College of Computing, Engineering, and Construction at the University of North Florida (UNF). He conducted meteorological and oceanographic research for over 40 years, leading many efforts that have contributed significantly to improving the predictive state of the art for winds, waves, currents, surges, and coastal evolution caused by storms, along with improved methods for the quantification of risk which incorporate alleatory and epistemic uncertainty into a consistent physical framework for coastal hazards. He serves as a U.S. delegate to the United Nations' Joint World Meteorological Organization and Intergovernmental Oceanographic Commission (WMO-IOC) Technical Commission for Oceanography and Marine Meteorology in the area of climate effects and the ocean and is the co-chair of the UN Coastal Inundation and Flooding Demonstration Project. Before coming to UNF, Dr. Resio served as Senior Scientist for the Engineer Research and Development Center's Coastal and Hydraulic Lab.

Victoria Sankovich: Victoria Sankovich is a Meteorologist in the Flood Hydrology and Consequences Group in the Technical Service Center (TSC) of the Bureau of Reclamation in Denver, Colorado. She has served in the TSC for over three years, conducting extreme precipitation and storm research and project work. Ms. Sankovich previously studied cloud/precipitation regimes as predicted by numerical weather models using statistical methods. Further, she has operational experience as a meteorologist for the U.S. Antarctic Program. This unique opportunity provided a challenging experience to work directly with weather instrumentation, including LiDAR, sunshine recorders, snow stakes, and rawinsondes. Her principal responsibility is providing expert technical advice on hydrometeorological statistics, including L-moments, and extreme storm analysis to TSC management and staff concerning hydrologic issues for Reclamation and other facilities of the Department of the Interior.

Mel Schaefer: Mel Schaefer is a Surface Water Hydrologist with a Ph.D. in Civil Engineering from the University of Missouri-Rolla. He has over 35 years of experience in hydrologic applications, primarily in analysis of extreme storms and floods, and was head of the Washington State Dam Safety program for 8 years. He has developed methodologies for conducting regional precipitation-frequency analyses and for stochastic simulation of the spatial and temporal characteristics of storms for use in rainfall-runoff modeling of extreme storms and floods for watersheds throughout the western U.S. and British Columbia. He is co-developer of the Stochastic Event Flood Model (SEFM) for conducting simulations of extreme floods. He has conducted studies for extreme floods and provided technical assistance to the U.S. Department of the Interior's Bureau of Reclamation, the U.S. Army Corps of Engineers, BC Hydro, and other hydropower utilities on more than 20 dams to develop probabilistic loadings for extreme floods for use in risk analyses for high-consequence dams.

Ray Schneider: Mr. Schneider is a Fellow engineer for the Westinghouse Electric Company. He has more than forty years of experience in the area of Light Water Reactor design and safety analysis, with emphasis on risk-informed probabilistic safety assessment (PSA) applications and severe accident analysis. Over the past several years, Mr. Schneider has represented the Pressurized Water Reactor Owners Group (PWROG) as a member of several Nuclear Energy Institute (NEI) task forces, including the NEI Risk-Based Task Force, Risk-Informed Technical Specifications Task Force, Fire Maintenance Rule Task Force (MRTF-Fire) and the NEI FLEX task force supporting the development of the Programmatic Controls section. Mr. Schneider is currently the technical lead for the PWROG supporting the NEI Fukushima Flooding Task Force. Mr. Schneider also serves on several standards-development committees, including the American Nuclear Society/American Society of Mechanical Engineers (ANS/ASME) Joint Committee on Nuclear Risk Management, ANS/ASME Level 2 Probabilistic Risk Assessment (PRA) Standards working group, and the ANS 2.8 flood hazard development working group. He regularly supports Pressurized Water Reactor Owner's Groups activities and individual member utilities in Level 1 PSA methods and applications development and activities related to Severe Accident Management, and has been the lead individual responsible for over 60 owner's group and individual utility tasks involving PSA methods development, PSA quality improvement, maintenance rule implementation, and risk-informed applications. Mr. Schneider has authored or co-authored over 30 technical publications and is the recipient of a number of awards, including the ASME Engineer of the Year Award for the Hartford Chapter and several division-level George Westinghouse Signature Awards.

Timo Schweckendiek: Timo Schweckendiek is a researcher and consultant at Deltares since 2006. His background is hydraulic and geotechnical engineering; he obtained his M.Sc. in Civil Engineering from Delft University of Technology, with which he is affiliated now through his Ph.D. research on risk-based site investigation for levees. His core fields of expertise are flood risk analysis and management and reliability analysis of flood defenses.

He is member of the Dutch Expertise Network of Flood Protection, an independent advisory body to the Dutch government on flood defenses and risk issues. Furthermore, he is a Netherlands delegate to Technical Committee TC304 on "Risk Management in Engineering Practice" of the International Society of Soil Mechanics and Geotechnical Engineering. Recently, in November 2012, Mr. Schweckendiek was co-organizer of the FLOODrisk 2012 conference held in Rotterdam.

Jery Stedinger: Dr. Stedinger is a Professor of Environmental and Water Resources Systems Engineering in the School of Civil and Environmental Engineering at Cornell University. Dr. Stedinger received a B.A. from the University of California–Berkeley in 1972 and a Ph.D. in Environmental Systems Engineering from Harvard in 1977. Since that time he has been a professor in Cornell's School of Civil and Environmental Engineering. Dr. Stedinger's research has focused on statistical and risk issues in hydrology and the operation of water resource systems. Projects have addressed flood frequency analysis, including the use of historical and paleoflood data, dam safety, regional hydrologic regression analyses, risk and uncertainty analysis of flood-risk-reduction projects, climate change modeling, water resource system simulation, and efficient hydropower system operation and system design.

John Stevenson: Dr. Stevenson is a professional engineer with many years of engineering experience, including the development of structural and mechanical construction criteria, codes, and standards used in construction of nuclear power plants and other hazardous civil and mechanical structures, distribution systems, and components subjected to postulated extreme environmental and accident loads as a design basis.

He is, or has been since 1970, an active member or chairman of the American Nuclear Society, American Society of Civil Engineers, American Society of Mechanical Engineers (ASME), American Concrete Institute, American Institute of Steel Construction, and the International Atomic Energy Agency (IAEA), where since 1975 he has actively participated in the development of a large number of International Safety Guides, Tech Docs., and Safety Reports concerning siting and civil and mechanical design of nuclear facilities. He currently serves as Chairman of the Technical Advisory Committee to the IAEA International Seismic Safety Center. He is a member of the ASME Boiler and Pressure Vessel (B&PV) Code Committee (for Section III, "Nuclear Components") and various Section III Working Groups, Subgroups, and Subcommittees (including as a former Chairman of the Subcommittee for ASME B&PV Code, Section III, Division 3, "Transportation and Storage Containments for Nuclear Material and Waste"). He is a former member of the ASME Board of Nuclear Codes and Standards and is a fellow of ASME. He is also a member of the American Nuclear Society (ANS) Nuclear Facilities Standards Committee and is or has been a member or Chairman of the ANS 2.3, 2.8, 2.10, 2.14, 2.26, 2.31, and 58.16 Standards Working Groups who are or have been in the process of developing standards for seismic, wind, and flood design and safety categorization of structures, systems, and components located in nuclear facilities.

His previous work assignments have included Manager of Structural Design and Balance of Plant Engineering Standardization for a major nuclear steam system supplier, Manager of Corporate Quality Assurance for a major architectural/engineering firm, and Founder and President of a consulting engineering firm with five offices (two in the U.S. and three in Eastern Europe) dedicated to safe design of new and existing nuclear facilities. He is currently a Consulting Engineer providing outside expert consulting services to various organizations involved in the construction of nuclear safety-related facilities.

Dr. Stevenson received his B.S. in Civil Engineering from the Virginia Military Institute in 1954, M.S. in Civil Engineering from the Case Institute of Technology in 1962, and Ph.D. in Civil Engineering from Case Western Reserve University in 1968.

Hong Kie Thio: Hong Kie Thio is Principal Seismologist at the Los Angeles office of URS Corporation. He joined the company in 1995 after receiving his Ph.D. in Geophysics at Caltech and has worked on numerous projects, including earthquake source studies, realtime earthquake analysis, nuclear monitoring, seismic hazard analysis, and tsunami hazard analysis. He is currently serving on the American Society of Civil Engineers (ASCE) 7 subcommittee on tsunami loads.

Wilbert O. Thomas, Jr.: Mr. Thomas is a Senior Technical Consultant with Michael Baker Jr., Inc. in Manassas, Virginia. He reviews and performs hydrologic analyses for flood insurance studies and participates in special projects sponsored by the Federal Emergency Management Agency and other clients such as the Maryland State Highway Administration. Mr. Thomas worked for the U.S. Geological Survey for 30 years and has worked for Michael Baker, Jr., Inc., since April 1995. Mr. Thomas is a registered professional hydrologist with over 47 years of specialized experience in conducting water resources projects and analyzing water resources data. Mr. Thomas is the author of more than 75 papers and abstracts on a variety of surface water hydrologic topics. As a U.S. Geological Survey employee, Mr. Thomas participated in the development of Bulletin 17B, "Guidelines for Determining Flood Flow Frequency," published in 1982 and still used by all Federal agencies for flood frequency analysis for gauged streams. Mr. Thomas is the current chair of the Hydrologic Frequency Analysis Work Group (http://acwi.gov/hydrology/Frequency/) that is evaluating revisions to Bulletin 17B.

Tony Wahl: Tony Wahl is a Hydraulic Engineer with 23 years' experience in the Bureau of Reclamation's Hydraulics Laboratory at Denver, Colorado, working in the areas of flow measurement and canal control, turbine hydraulics, fish screening, dam breach modeling, and laboratory modeling and field testing of hydraulic structures. Since 1995 he has been engaged in research related to embankment dam breach modeling. Since about 2004 he has been the technical coordinator for an international collaboration through the Centre for Energy Advancement through Technological Innovation's Dam Safety Interest Group that is focused on developing better methods for modeling embankment dam erosion and breach processes. He has led studies to evaluate competing methods for quantifying the erodibility of cohesive soils, a key input to newer dam erosion and breach models. Mr. Wahl's recent research also includes physical scale modeling and numerical modeling of breach processes associated with canal embankments. Mr. Wahl is a registered professional engineer and earned a B.S. and M.S. in Civil Engineering from Colorado State University. In 2005, he was selected as the Bureau of Reclamation's Federal Engineer of the Year. Mr. Wahl is the author of more than 90 technical papers and is presently serving on American Society of Civil Engineers task committees studying flow measurement at canal gates and the uncertainties and errors associated with hydraulic measurements and experimentation.

Ty V. Wamsley: Dr. Ty Wamsley is Chief of the Storm and Flood Protection Division in the Coastal and Hydraulics Laboratory at the U.S. Army Engineer Research and Development Center in Vicksburg, MS. He leads a team of engineers and scientists in the execution of research and development and engineering support services in the areas of storm and flood risk-reduction project design and maintenance; sediment management; groundwater and watershed engineering and management; operation and maintenance of navigation projects; and regional-scale databases and information systems. He has received a Department of the Army Superior Civilian Service Award for his leadership and technical contribution to the Corps' flood and storm damage reduction mission.

Yong Wei: Dr. Yong Wei is a senior research scientist in tsunami modeling, coastal hazard mitigation, and geophysical data analysis. He has more than 10 years of research experience on tsunami modeling, methodology, and associated geophysics. He has developed extensive skills and proficiencies in various computer models dealing with the hydrodynamics of long waves. He also specializes in analysis of the tsunami source mechanism from seismic data and ocean observations. He joined Pacific Marine Environmental Laboratory and University of Washington in 2006. He is currently the leading scientist in the National Oceanic and Atmospheric Administration's tsunami modeling team in a multiyear effort to develop high-resolution tsunami inundation models for U.S. coasts.

Daniel Wright: Daniel Wright is a 4th-year Ph.D. candidate in Environmental Engineering and Water Resources at Princeton University, focusing on flood risk estimation in urban environments. At Princeton, he is also a participant in the Science, Technology, and Environmental Policy certificate program, looking at climate and land use impacts on flood hazards. He has Bachelor's and Master's degrees in Civil and Environmental Engineering from the University of Michigan, where he focused on water resources engineering. From 2006–2008 Daniel served as a Regional Sanitation Engineer in the Peace Corps in Monteagudo, Bolivia, and later as a hydraulic engineer at an engineering firm in Concepción, Chile, developing low-impact hydroelectric power projects. He is interested in the intersection of engineering, policy, and development in water resources and climate adaptation.

NRC FORM 335 (12-2010) NRCMD 3.7	U.S. NUCLEAR REGULATORY COMMISSION	1. REPORT NUMBER (Assigned by NRC, Add Vol., Supp., Rev., and Addendum Numbers, if any.)
	BIBLIOGRAPHIC DATA SHEET *(See instructions on the reverse)*	NUREG/CP-0302

2. TITLE AND SUBTITLE	3. DATE REPORT PUBLISHED	
Proceedings of the Workshop on Probabilistic Flood Hazard Assessment (PFHA), U.S. NRC Headquarters, Rockville, MD, January 29 - 31, 2013	MONTH	YEAR
	October	2013
	4. FIN OR GRANT NUMBER	

5. AUTHOR(S)	6. TYPE OF REPORT
compiled by Thomas J. Nicholson and Wendy Reed, U.S. NRC, Office of Nuclear Regulatory Research, Division of Risk Analysis	Workshop Proceedings
	7. PERIOD COVERED (Inclusive Dates)
	January 29 - 31, 2013

8. PERFORMING ORGANIZATION - NAME AND ADDRESS (If NRC, provide Division, Office or Region, U. S. Nuclear Regulatory Commission, and mailing address; if contractor, provide name and mailing address.)

Division of Risk Analysis
Office of Nuclear Regulatory Research
U.S. Nuclear Regulatory Commission
Washington, DC 20555-0001

9. SPONSORING ORGANIZATION - NAME AND ADDRESS (If NRC, type "Same as above", if contractor, provide NRC Division, Office or Region, U. S. Nuclear Regulatory Commission, and mailing address.)

Same as above

10. SUPPLEMENTARY NOTES

Federal interagency workshop focusing on probabilistic flood hazard assessment

11. ABSTRACT (200 words or less)

NRC's Offices of Nuclear Regulatory Research, Nuclear Reactor Regulation and New Reactors organized this Workshop on Probabilistic Flood Hazard Assessment (PFHA). The workshop was held January 29–31, 2013 at NRC headquarters. The workshop was coordinated with Federal agency partners: the U.S. Department of Energy; U.S. Department of the Interior's Bureau of Reclamation and U.S. Geological Survey; U.S. Army Corps of Engineers; and the Federal Energy Regulatory Commission. The research workshop was devoted to the sharing of information on PFHA for extreme events (ice, annual exceedance probabilities much less than 2.0E–3 per year). The topics included: Federal agencies' interests and needs in PFHA; State-of-the-Practice in Identifying Extreme Flood Hazards; Extreme Precipitation Events; Flood-Induced Dam and Levee Failures; Tsunami Flooding; Riverine Flooding; Extreme Storm Surges for Coastal Areas; and Combined Events Flooding. The workshop objectives included to: (1) assess, discuss, and inform workshop participants on the state of the practice for extreme flood assessments within a risk context; (2) facilitate the sharing of information between both Federal agencies and other interested parties to bridge the current state of knowledge between extreme flood assessments and risk assessments of critical infrastructures; (3) seek ideas and insights on possible ways to develop a PFHA for use in probabilistic risk assessments; (4) identify potential components of flood-causing mechanisms that lend themselves to probabilistic analysis and warrant collaborative study; and (5) establish realistic plans for coordination of PFHA research studies. Observations and insights provided during the session presentations and subsequent panel discussions, that followed, were documented by the panel reporters and are included in the proceedings.

12. KEY WORDS/DESCRIPTORS (List words or phrases that will assist researchers in locating the report.)		13. AVAILABILITY STATEMENT
combined events for flood assessments	risk-informed decisionmaking	unlimited
dam and levee safety and failures	riverine flooding	14. SECURITY CLASSIFICATION
Expert Elicitation Flood Hazard Assessment	Senior Seismic Hazard Analysis Committee	*(This Page)*
extreme precipitation events	tsunami flooding	unclassified
extreme storm surge for coastal areas	watershed modeling	*(This Report)*
flood frequency		unclassified
paleoflood hydrology		15. NUMBER OF PAGES
probabilistic flood hazard assessment		
probabilistic tsunami hazard assessment		16. PRICE

NRC FORM 335 (12-2010)

United States Nuclear Regulatory Commission
October 2013

NUREG/CP-0302

Proceedings of the Workshop on Probabilistic
Flood Hazard Assessment (PFHA)

October 2013

www.ingramcontent.com/pod-product-compliance
Lightning Source LLC
Chambersburg PA
CBHW080239180526
45167CB00006B/2340